Monographs in Mathematical Economics

Volume 9

This series, consisting of research monographs and advanced textbooks for graduate students, is designed to bring together mathematicians interested in obtaining challenging new stimuli from economic theories and economists seeking effective mathematical tools for their research.

The scope of the series includes but is not limited to:

- Economic theories in various fields based on rigorous mathematical reasoning
- Mathematical methods (e.g., analysis, algebra, geometry, probability) motivated by economic theories
- Mathematical results of potential relevance to economic theory
- Historical study of mathematical economics

Comparable, existing monographs series are mainly organized from the viewpoint of "users" of mathematics, and are thus of limited interest to mathematicians. This series in contrast aims at genuine interaction between economists and mathematicians.

Most economic phenomena are described by (1) optimizing behaviors of agents (consumers, firms, governments, etc.) and (2) equilibria generated through the interactions of these agents. Consequently, the most basic mathematical subjects for economics fall into two categories:

- Optimization theory, static and dynamic. Basic linear / nonlinear programming, calculus of variations, optimal control, dynamic programming etc. provide key mathematical tools. Modern developments in convex analysis, nonlinear functional analysis, set-valued analysis, and non-smooth analysis form indispensable foundations.
- Equilibrium theory. Existence in finite- / infinite-dimensional settings, characterization, computational algorithms, dynamic adjustment processes, correspondence with economic structures.

Some fixed-point theorems and variational inequalities were created to solve the existence problem of equilibria. The stability theory and the viability theory of differential equations play important roles in the dynamic aspects of the equilibrium theory. Further, differential topology is a basic tool for the geometric analysis of equilibrium manifolds.

Economic phenomena significant in the real world often have a dynamic character. For instance, business cycles and movements of asset prices are governed by dynamic laws expressed in terms of differential or difference equations. Nonlinear analysis, harmonic analysis, and stochastic calculus play significant roles in solving dynamic economic problems. Econometric methods, including time series analysis and forecasting, also require sophisticated mathematical tools. Finally, significant developments in game theory and interaction with mathematical logic should be mentioned.

All of these topics and phenomena come within the scope of this series, aimed at cooperation between economists and mathematicians and the development of their respective disciplines.

Takashi Hayashi

Decision Theory

Takashi Hayashi
Adam Smith Business School
University of Glasgow
Glasgow, UK

ISSN 2364-8279 ISSN 2364-8287 (electronic)
Monographs in Mathematical Economics
ISBN 978-981-95-2199-9 ISBN 978-981-95-2200-2 (eBook)
https://doi.org/10.1007/978-981-95-2200-2

This Springer imprint is published by the registered company Springer Nature Singapore Pte Ltd.
The registered company address is: 152 Beach Road, #21-01/04 Gateway East, Singapore 189721, Singapore

Preface

This is an introductory textbook on decision theory that is based on an axiomatic approach and intended for graduate students or educated non-professionals. Because it is meant to be "introductory," I have put priority on letting the reader get used to how the axiomatic approach works, through following the mathematical presentation closely line by line. In other words, this is an introduction to the method, not an introduction to the topics. Because of my limited capacity, this choice led me to sacrifice fully covering up-to-date development of the theory in terms of behavioral broadness. I apologize in advance.

I decided to write a textbook with such a style nevertheless, because very few schools offer a second-year graduate course on axiomatic decision theory and students who don't go to those schools can get exposed to only "take-away results" at most unless they can read academic theory papers by themselves. In my and my predecessors' age, notable textbooks in axiomatic decision theory, such as Fishburn [35] and Kreps [72], have become classics that I rely on much in the current book as well. I hope there is room for publishing another book, as these books are getting less available and also they need to be updated with developments afterwards. Although even mine is not entirely recent, I tried to cover developments up to the 2010s that are necessary and sufficient for the reader to learn the "basic moves" of an axiomatic approach.

An axiom has two roles in decision theory. First, when your interest is a *normative* one, asking "how decision making *should* be," then the role of the axiom is to clearly spell out a criterion of "rationality." For example, the axiom saying "if you choose X over Y and choose Y over Z then you choose X over Z," which is called transitivity, extracts an aspect of "rationality" as consistency.

Second, when your interest is a *positive* one about "how *does* a human make decisions?" the role of the axiom is to describe a property that is directly testable through observation of choices. There, the transitivity condition "if you choose X over Y and choose Y over Z then you choose X over Z" is supposed to describe a property that is supposed to be met by human's choice behavior, and it is directly testable by observing choices between X and Y, between Y and Z, and between X and Z.

After presenting such a set of axioms, axiomatic analysis presents its result in the form of a representation theorem, a statement like

> a decision rule satisfies a certain set of axioms if and only if it is represented as the maximization of a certain function.

Note that "a certain function" is something we "back up" from observed choices. For, we cannot dig in our brain and physically pick up and show "a certain function," although I'm aware that such an attempt has already a long history. In this book let me still maintain a conservative view about it.

While the role of axioms is relatively literal and clear in a normative argument, the role of axioms in a positive argument will need some examination. To compare, think of an attempting to start from a plausible functional form for the objective and describing a real choice as the maximization solution of it. The problem there is that we cannot know which property of the function is wrong when the solution does not fit observation or data. This leads us to repeat manipulating the functional form at random or repeat adding parameters at random. Although it is a matter of aesthetics of research, basically the axiomatic people hate this. Hence they try to make the theory directly testable/falsifiable through observation, as much as possible, by means of writing down a property of choice behavior in the form of an axiom, an independent statement.

Honestly, at the level of empirical analysis, axioms are mostly falsified rather than falsifiable. This is natural, I would say, because it is always difficult to find a formal and exact law in human choices. Also, when choice objects change in a continuous manner, verifying an axiom typically requires infinite data.

The reason why we think of an axiom in positive arguments nevertheless would be that it keeps us conscious of what aspect of reality we analysts abstract away, and it forces us to be clear about what normative problem we will face when it is indeed violated by real choices.

The same thing can be said about normative arguments, though. It happens that an implication of some decision criterion contradicts our "ethical intuition." Then, we will end up with manipulating a functional form of decision criterion at random if we do not write down such "ethical intuition" in the form of a set of axioms.

Keeping the above in mind, this book intends to help the readers to get used to the "basic moves" of an axiomatic method, by going through basic topics in decision theory with proofs. Acquisition of proof techniques is critically important in understanding the role of axioms. Because of this policy and because of my capacity, honestly saying again, the topics selected are not very comprehensive at the research level nowadays and also are biased. Once you get used to the "basic moves," however, I think it is nevertheless helpful in catching up with recent research as well.

The "basic moves" here are

1. Determine the primitive of the model, where the primitives refer to objects or things which are regarded as directly observable. Which are directly observable is not an obvious question by itself, but what is important is that once we determine

the primitives we do not change them throughout the analysis, and we do the analysis under that restriction.
2. Write down axioms, propositions which are written only with the primitives and are directly testable through observations.
3. Draw the implication of the axioms in the form of a representation theorem.

About Mathematics While this book is intended to be an "introduction," since it is intended for graduate students or educated non-professionals, I did not try to dispense with mathematical expositions. Most of the materials can be understood just with set-theoretic logic and basic knowledge of topology on finite-dimensional Euclidian spaces, though. Even infinite-dimensional objects could be mostly understood by an analogy of finite-dimensional vectors. For example, you could imagine a probability distribution as a non-negative vector whose entries add up to one. Otherwise, I would recommend you to consult a reference book on general topology, real analysis, and measure theory such as Aliprantis and Border [3], or you might want to refresh basic knowledge by going back to a classic textbook such as Rudin [89]. Just in case, I have put $*$ on the sections which require knowledge on measure theory.

Acknowledgments This book was originally published in Japanese as a part of the *Mathematical Economics Series*, by Chisen Shokan Publisher in 2020, while I added significant amounts of material in this English version. I thank Toru Maruyama, Chiaki Hara, Norio Takeoka, Youichiro Higashi, and Mitsuo Koyama for comments and encouragement regarding the Japanese version. I thank Toru Maruyama, the editor of *Monograph in Mathematical Economics*, and Masayuki Nakamura of Springer for providing me the opportunity to write an English version. I thank three anonymous reviewers and Georgios Gerasimou for helpful comments and suggestions on the English version. Finally, I thank Larry Epstein, who I learned the axiomatic decision theory from.

Glasgow, UK Takashi Hayashi

Contents

Chapter 1
Preference and Its Representation

1.1 Preference Relation

Let X denote the set of all the possible choice alternatives. At this point the set is abstract and its elements could be anything.

A binary relation, denoted by $\succsim$, describing preference over X is given as a subset of $X \times X$, that is, $\succsim \subset X \times X$.

For an arbitrary pair of elements of X such as $(x, y) \in X \times X$, when $(x, y) \in \succsim$ holds, we read it as "x is at least as good as y."

Now we switch to our familiar way of writing a binary relation and write

$$x \succsim y$$

instead of $(x, y) \in \succsim$ hereafter.

Then we write

$$x \succ y$$

when $x \succsim y$ holds and $y \succsim x$ does not, and read it as "x is (strictly) better than y," and we write

$$x \sim y$$

when both $x \succsim y$ and $y \succsim x$ hold, and read it as "x and y are equally preferable" or "the decision maker is indifferent between x and y."

In normative arguments such a binary relation is naturally supposed to rank choice alternatives according to their desirability.

When it comes to descriptive/positive arguments, on the other hand, can we directly observe such a preference relation? Here we maintain the conservative standpoint that we cannot say "Hey, this is your preference" by directly cutting

T. Hayashi, *Decision Theory*, Monographs in Mathematical Economics 9,
https://doi.org/10.1007/978-981-95-2200-2_1

into your brain. Hence what we mean here is that we DEFINE $x \succsim y$ by the fact that you choose x when the pair x and y are available to choose (which does not yet exclude that you choose y as well). Likewise, we DEFINE $x \succ y$ by the fact that you choose x and not y when the pair x and y are available to choose. Such a standpoint is called the **revealed preference** approach.

The real-life choice problems are not necessarily binary. Hence there must be certain consistency conditions on choice behaviors/data so that they are explained as maximization of a binary relation. Let me relegate the argument on this to the next chapter and proceed for now by assuming that a binary relation exists and it is observable.

We introduce two standard axioms on preference relations. One is completeness.

Completeness: For all $x, y \in X$, either $x \succsim y$ or $y \succsim x$ holds.

Note that "either ... or" in the above statement does not exclude the case that both $x \succsim y$ and $y \succsim x$ hold, which is the case of indifference.

Intuition of completeness is that the decision maker can rank between any two alternatives, including the case of indifference that they are equally preferable.

As it may sound trivial, let us see the significance of it by looking at the following example.

Example 1.1 Let $X = \mathbb{R}^n$ and define $\succsim$ by

$x \succsim y$ if and **only if** $x_k \geqq y_k$ for all $k = 1, \cdots, n$.

This relation is incomplete, since x and y are not comparable for example when $x_1 > y_1$ and $x_2 < y_2$.

Under completeness, the following notation will make sense:

- Write $x \succ y$ if $x \succsim y$ and $y \not\succsim x$, which we read as "x is better than y."
- Write $x \sim y$ if $x \succsim y$ and $y \succsim x$, which we read as "the decision maker is indifferent between x and y," or "x is equally preferable to y."

The other standard axiom is transitivity, which says that preference does not cycle.

Transitivity: For all $x, y, z \in X$, if $x \succsim y$ and $y \succsim z$, then $x \succsim z$.

Here is an example of intransitive preference.

Example 1.2 Let $X = \mathbb{R}^n$ and define $\succsim$ by

$$x \succsim y \iff \max_{k=1,\cdots,n} \{\max\{x_k, y_k\} - x_k\} \leq \max_{k=1,\cdots,n} \{\max\{x_k, y_k\} - y_k\}$$

This relation is complete but intransitive. To see this, consider for example

$$x = (3, 1, 0),\ y = (0, 3, 1),\ z = (1, 0, 3),$$

then we have

$$x \succ y \succ z \succ x.$$

1.2 Representation of Preference

In economics, we analyze an actor's preference by means of numerically representing it with a so-called utility function. Since this wording is misleading as it gives an impression as if a utility exists as a "substance," we avoid this as much as possible in this book.

Definition 1.1 A function $u : X \to \mathbb{R}$ is said to represent preference $\succsim$ if for all $x, y \in X$

$$x \succsim y \text{ if and only if } u(x) \geqq u(y).$$

From the definition, it should also follow that

$$x \succ y \text{ if and only if } u(x) > u(y)$$

and

$$x \sim y \text{ if and only if } u(x) = u(y).$$

Before talking about whether a preference relation admits representation by a function, first let us see what meaning such a function has.

Proposition 1.1 *Suppose both of u and v represent the same preference relation $\succsim$, then there is a monotone increasing function $f : u(X) \to \mathbb{R}$ such that $v = f \circ u$. The converse is true as well.*

Proof $\Longrightarrow$ part: define a function $f : u(X) \to \mathbb{R}$ by $f(w) = v(x)$ for each $w \in u(X)$, by taking $x \in X$ such that $u(x) = w$. Since both u, v represent the same preference relation, if $u(x) = u(y)$ it must be $v(x) = v(y)$, hence the definition of f does not depend on the choice of x. Also, if $w > w'$, by taking $x, y \in X$ with $u(x) = w$ and $u(y) = w'$, we see $x \succ y$ since u is a representation of $\succsim$. Thus $f(w) = f(u(x)) = v(x) > v(y) = f(u(y)) = f(w')$ since v is a representation of $\succsim$, which shows f is increasing.

$\Longleftarrow$ part: Start with a function u which represents $\succsim$ and consider any monotone increasing function $f : u(X) \to \mathbb{R}$, and define a function $v = f \circ u$. Then, since f

is monotone increasing,

$$\begin{aligned} v(x) \geqq v(y) &\Longleftrightarrow f(u(x)) \geqq f(u(y)) \\ &\Longleftrightarrow u(x) \geqq u(y) \\ &\Longleftrightarrow x \succsim y \end{aligned}$$

for all $x, y \in X$. Hence v is also a representation of $\succsim$. □

This shows that if a preference relation has "some" representation one can generate arbitrarily many representations by means of taking an arbitrary monotone transformation. And we do not have a theory about which representation is the "right" one among such arbitrarily many ones representing the same preference ranking. For example, suppose we double a representation u and obtain $v = 2u$. If we use the latter instead of the former, is the decision maker "doubly happy?" This statement has no meaning, as far as we look at choice behavior of individuals. Whether you use u or v, the implied choice behavior is the same.

We call a numerical expression **ordinal** when it has a meaning only as representation of a ranking and has no quantitative meaning beyond that. On the other hand, we call a mathematical object **cardinal** when it has a quantitative meaning and it can be "measured" in a suitable sense. Any representation of preference here is thus ordinal.

For example, in the context of economics in which a vector $x = (x_1, x_2)$ is understood as having x_1 units of Good 1 and x_2 units of Good 2, consider preference represented by a function

$$u(x) = ax_1 + bx_2.$$

Double it and obtain

$$v(x) = 2ax_1 + 2bx_2$$

or take exponential transformation and obtain

$$w(x) = e^{ax_1 + bx_2},$$

then these are ordinal representations of the same preference and there is no economic meaning in difference among them, in the sense that they imply the same choice behavior.

On the other hand, when we plot equally preferable points in the above on a two-dimensional plane they form a straight line with slope $-\frac{a}{b}$. In economics, we call it an indifference curve. The above example is the case that indifference curves are parallel and straight, and that the decision maker is willing to sacrifice $\frac{a}{b}$ units of Good 2 in order to get one extra unit of Good 1. Such a rate is called the marginal rate of substitution and it is measurable and hence cardinal.

When does preference admit representation? First let us consider the case that X is a countable set.

Proposition 1.2 *Suppose X is countable. Then, if $\succsim$ is complete and transitive, there is a function $u : X \to \mathbb{R}$ that represents it, and it can be taken so that $u(X) \subset \mathbb{Q} \cap (0, 1)$. The converse is true as well.*

Proof Necessity of the axiom is straightforward, so we prove sufficiency. First, it is obvious when X is a finite set. For example, one can define

$$u(x) = \frac{|\{y \in X : x \succsim y\}|}{|X| + 1}.$$

Thus we assume X is countably infinite, and denote it by

$$X = \{x_1, x_2, \cdots\}.$$

Since $\mathbb{Q} \cap (0, 1)$ is a countably infinite set, it is written as

$$\mathbb{Q} \cap (0, 1) = \{u_1, u_2, \cdots\} = \{u_k\}_{k \in \mathbb{N}}.$$

Now define $u : X \to \mathbb{Q} \cap (0, 1)$ as follows:

- $u(x_1) = u_1$.
- For x_2,
  - If $x_2 \sim x_1$, then set $u(x_2) = u(x_1) = u_1$.
  - If $x_2 \succ x_1$, then set $u(x_2)$ to be the first number in $\{u_2, u_3, \cdots\}$ that is greater than u_1.
  - If $x_2 \prec x_1$, then set $u(x_2)$ to be the first number in $\{u_2, u_3, \cdots\}$ that is smaller than u_1.
- Inductively, for x_k:
  - If $x_k \sim x_l$ for some $l = 1, \cdots, k-1$, then set $u(x_k) = u(x_l)$.
  - If $x_k \succ x_l$ for all $l = 1, \cdots, k-1$, then set $u(x_k)$ to be the first number in $\{u_1, u_2, u_3, \cdots\} \setminus \{u(x_1), u(x_2), \cdots, u(x_{k-1})\}$ that is greater than all of $u(x_1), u(x_2), \cdots, u(x_{k-1})$.
  - If $x_k \prec x_l$ for all $l = 1, \cdots, k-1$, then set $u(x_k)$ to be the first number in $\{u_1, u_2, u_3, \cdots\} \setminus \{u(x_1), u(x_2), \cdots, u(x_{k-1})\}$ that is smaller than all of $u(x_1), u(x_2), \cdots, u(x_{k-1})$.
  - If $x_l \succ x_k \succ x_m$ for some $l, m = 1, \cdots, k-1$, and such l and m are taken so that there is no $n = 1, \cdots, k-1$ such that $x_l \succ x_n \succ x_m$, then set $u(x_k)$ to be the first number in the intersection of $\{u_1, u_2, u_3, \cdots\} \setminus \{u(x_1), u(x_2), \cdots, u(x_{k-1})\}$ and $(u(x_l), u(x_m))$.

By definition, $u(x) \geqq u(y)$ holds if and only if $x \succsim y$. □

Now we extend the representation to possibly uncountable sets, like $\mathbb{R}^n$, by following the classic results such as Debreu [22].

Definition 1.2 $(X, \succsim)$ has an order-dense countable subset if there is a countable subset $Y \subset X$ and for all $x, y \in X$ with $x \succ y$ there is $a \in Y$ such that $x \succ a \succ y$.

Proposition 1.3 *Preference $\succsim$ defined over X satisfies completeness and transitivity, and $(X, \succsim)$ has an order-dense countable subset if and only if there is a function $u : X \to \mathbb{R}$ which represents $\succsim$.*

Proof It is trivial if $x \sim y$ holds for all $x, y \in X$, hence we assume $x \succ y$ for some $x, y \in X$.

Denote an order-dense countable subset of X with regard to $\succsim$ by Y, then by Lemma 1.2 there is a function $u : Y \to \mathbb{Q} \cap (0, 1)$ which represents $\succsim$ restricted to Y. Now extend this to the entire X by

$$u(x) = \sup\{u(z) : x \succ z,\ z \in Y\}.$$

We show that u indeed represents $\succsim$ over X. For arbitrary $x, y \in X$, suppose $x \succsim y$. Then, for all $z \in Y$, if $y \succ z$ then it follows that $x \succ z$ because $x \succsim y \succ z$. Hence $\{z \in Y : x \succ z\} \supset \{z \in Y : y \succ z\}$. By the definition of an upper-bound, for all $z \in Y$ with $x \succ z$, $\sup\{u(z) : x \succ z,\ z \in Y\} \geqq u(z)$. Hence from the inclusion relation, for all $z \in Y$ with $y \succ z$, $\sup\{u(z) : x \succ z,\ z \in Y\} \geqq u(z)$. Hence $\sup\{u(z) : x \succ z,\ z \in Y\}$ is an upper-bound of the set $\{u(z) : y \succ z,\ z \in Y\}$. Thus, from the definition of the least upper-bound we have $\sup\{u(z) : x \succ z,\ z \in Y\} \geqq \sup\{u(z) : y \succ z,\ z \in Y\}$, hence $u(x) \geqq u(y)$.

Also, suppose $x \succ y$. Then, by the order-dense property there is $a \in Y$ such that $x \succ a \succ y$, and again by the order-dense property there is $b \in Y$ such that $x \succ a \succ b \succ y$. In the representation already obtained over Y, we have $u(a) > u(b)$. From the previous step, we already have $u(x) \geqq u(a)$ at least, hence $u(x) > u(b)$. On the other hand, for any $z \in Y$ with $y \succ z$, $b \succ y \succ z$, which implies $b \succ z$. Then, since $u(b) > u(z)$ in the representation already obtained over Y, $u(b)$ is an upper-bound of the set $\{u(z) : y \succ z,\ z \in Y\}$. Therefore, by the definition of the least upper-bound, $u(b) \geqq \sup\{u(z) : y \succ z,\ z \in Y\} = u(y)$. Summing up, since $u(x) > u(b) \geqq u(y)$, we obtain $u(x) > u(y)$. □

Finally, we consider representability of preference when the set of alternatives has a topological structure.

Continuity: The set $\{(x, y) \in X \times X : x \succsim y\}$ is a closed subset of $X \times X$.

Proposition 1.4 *Let X be a separable and connected topological space. Then, preference $\succsim$ defined over X satisfies completeness, transitivity and continuity if and only if there is a continuous function $u : X \to \mathbb{R}$ which represents $\succsim$.*

Proof Since X is a separable topological space, it has a topologically dense countable subset, which we denote by Y. We show that this Y is an order-dense subset of X with regard to $\succsim$.

First, for arbitrary $x, y \in X$, suppose $x \succ y$. Then, by transitivity it must hold that $\{z \in X : z \succsim x\} \cap \{z \in X : y \succsim z\} = \emptyset$.

Now suppose $\{z \in X : x \succ z\} \cap \{z \in X : z \succ y\} = \emptyset$, then by completeness we must have $\{z \in X : z \succsim x\} \cup \{z \in X : y \succsim z\} = X$, but this means that X is partitioned to two mutually disjoint non-empty closed subsets, which contradicts the connectedness of X.

Hence $\{z \in X : x \succ z\} \cap \{z \in X : z \succ y\} \neq \emptyset$, and because this intersection is an open set and Y is topologically dense we have $Y \cap \{z \in X : x \succ z\} \cap \{z \in X : z \succ y\} \neq \emptyset$. □

Continuity is indispensable for existence of representation. We can see this from the following example.

Example 1.3 Define a preference ranking over $\mathbb{R}^2$ by

if $x_1 > y_1$, then $x \succ y$

and

if $x_1 = y_1$ and $x_2 > y_2$, then $x \succ y$

That is, a lexicographic preference. This is complete and transitive, but not continuous. For example, consider a sequence $(1 - 1/n, 1)$. Then we have

$$(1 - 1/n, 1) \prec (1, 0)$$

for all n, but in the limit we have

$$(1, 1) \succ (1, 0),$$

which is a violation of continuity.

Let us see that this lexicographic preference does not admit representation.

Suppose the preference admits representation $u : \mathbb{R}^2 \to \mathbb{R}$. Then, for every x_1, since $(x_1, 1) \succ (x_1, 0)$ it follows that $u(x_1, 1) > u(x_1, 0)$ from the definition of representation. Since the set of rational number is dense in the set of real numbers, we can take a rational number $r(x_1)$ such that

$$u(x_1, 1) > r(x_1) > u(x_1, 0).$$

Define such a function $r : \mathbb{R} \to \mathbb{Q}$, then it is a one-to-one function. For, if $x_1 > y_1$, since $u(x_1, 0) > u(y_1, 1)$, we have $u(x_1, 1) > r(x_1) > u(x_1, 0)$ and $u(y_1, 1) > r(y_1) > u(y_1, 0)$, which implies $r(x_1) > r(y_1)$.

However, the fact that $r : \mathbb{R} \to \mathbb{Q}$ is a one-to-one function contradicts the uncountability of the set of real numbers.

Chapter 2
Revealed Preference

In the previous chapter, we took a preference relation as a directly observable object. In descriptive arguments, however, we cannot cut into our brain and show "hey, this is preference," although I'm aware that there is a long history of such attempts. Here, let me take a conservative standpoint that all what we can observe directly is at most choice behavior.

Thus, preference is understood as something that can explain observed choice data as maximization of it, and has to be *backed up* from the data. Such a notion of preference is called revealed preference. Note that when we take preference as "directly observable" in descriptive arguments hereafter, we mean it as revealed preference.

2.1 Rationalizability by Maximization of a Weak Order

Let X be the set of all the alternatives, which may or may not be available to choose. We assume that X is a finite set. Let $\mathcal{B} \subset 2^X \setminus \{\emptyset\}$ be the family of **opportunity sets**, subsets of X which can be given to the decision maker as a set of available alternatives.

Observable choice data takes the form of a mapping $\varphi : \mathcal{B} \to 2^X \setminus \{\emptyset\}$, which maps each possible opportunity set $B \in \mathcal{B}$ into its non-empty subset $\varphi(B) \subset B$. This is a collection of data stating "the decision maker chooses $\varphi(B)$ when he is given a set of alternatives B" for each B, and called a **choice mapping**. Here we assume that φ is set-valued, not necessarily single-valued, as we intend to allow "ties." We will come to the case of single-valued choice mapping later.

Choice mapping φ itself may or may not explained as coming from maximization of some preference. When it is, we say that it is rationalizable.

T. Hayashi, *Decision Theory*, Monographs in Mathematical Economics 9,
https://doi.org/10.1007/978-981-95-2200-2_2

Definition 2.1 Choice mapping φ is rationalizable if there is a complete and transitive preference relation $\succsim$ over X such that

$$\varphi(B) = \varphi_{\succsim}(B) \equiv \{x \in B : \forall y \in B,\ x \succsim y\}$$

holds for all $B \in \mathcal{B}$.

Here are two necessary conditions for rationalizability.

Contraction: For all $B, C \in \mathcal{B}$, if $C \subset B$ then $\varphi(B) \cap C \subset \varphi(C)$.

The condition says that the world champion must be the champion in his country. Indeed, when there is a world ranking over all the players in the world, the world No.1 player will be the No.1 player in his country according to the same ranking.

One motivation for such a consistency condition is that if you fail to satisfy it you will face a dynamic inconsistency problem in the sense that discarding a choice alternative in an earlier stage changes how you choose from the remaining one in the later stages (Hammond [53]).

The second condition is concerned with cases that there are ties.

Expansion: For all $B, C \in \mathcal{B}$ and for all $x, y \in B$, if $B \subset C$ and $x, y \in \varphi(B)$ and $x \in \varphi(C)$, then $y \in \varphi(C)$.

The condition states, if there are multiple champions in a given country with ties and if one of them is in fact a world champion, then the others must be world champions as well with ties. Indeed, when there is a world ranking over all the players in the world, if there are several No.1 players in some country with ties and if one of them is in fact a No.1 player in the world then the others must be No.1 players in the world with ties.

Theorem 2.1 *Assume that the family of opportunity sets $\mathcal{B}$ consists of all the non-empty subsets of X. Then, choice mapping φ satisfies Contraction and Expansion if and only if it is rationalizable as the maximization of some complete and transitive preference relation $\succsim$.*

Proof We prove sufficiency of the axioms, as necessity is clear.

Suppose φ satisfies Completeness and Transitivity.

Define a binary relation $\succsim_\varphi$ by

$$x \succsim_\varphi y \iff x \in \varphi(\{x, y\}).$$

First we show that $\succsim_\varphi$ is complete and transitive. For all x, y, since $\varphi(\{x, y\}) \neq \emptyset$, either $x \in \varphi(\{x, y\})$ or $y \in \varphi(\{x, y\})$ must be true (both can be true of course). Hence either $x \succsim_\varphi y$ or $y \succsim_\varphi x$ must be true (both can be true of course, again). Hence $\succsim_\varphi$ is complete.

To show transitivity, suppose $x \succsim_\varphi y$ and $y \succsim_\varphi z$. By definition of $\succsim_\varphi$, we have $x \in \varphi(\{x, y\})$ and $y \in \varphi(\{y, z\})$. Now there are three cases.

(i) If $x \in \varphi(\{x, y, z\})$, then from Contraction it follows that $x \in \varphi(\{x, z\})$, meaning $x \succsim_\varphi z$.
(ii) If $y \in \varphi(\{x, y, z\})$, then from Contraction it follows that $y \in \varphi(\{x, y\})$. Since we have $x \in \varphi(\{x, y\})$ on the other hand, from Expansion it that follows that $x \in \varphi(\{x, y, z\})$. Hence it reduces to (i).
(iii) If $z \in \varphi(\{x, y, z\})$, then from Contraction it follows that $z \in \varphi(\{y, z\})$. Since we have $y \in \varphi(\{y, z\})$ on the other hand, from Expansion it follows that $y \in \varphi(\{x, y, z\})$. Hence it reduces to (ii).

Next we show that φ is indeed rationalizable as the maximization of $\succsim_\varphi$. Pick any $B \in \mathcal{B}$ and pick any $x \in \varphi(B)$. Then, from Contraction it follow $x \in \varphi(\{x, y\})$ for every $y \in B$, meaning $x \succsim_\varphi y$. Hence $\varphi(B) \subset \varphi_{\succsim_\varphi}(B)$.

Also, suppose $x \in B$ is such that $x \succsim_\varphi y$ holds for all $y \in B$. Then, since $\varphi(B) \neq \emptyset$ one can pick some $z \in \varphi(B)$. Then, from Contraction it follows $z \in \varphi(\{x, z\})$. On the other hand, since $x \succsim_\varphi z$ holds by assumption we have $x \in \varphi(\{x, z\})$. Hence, by Expansion we have $x \in \varphi(B)$. Thus we have shown $\varphi(B) \supset \varphi_{\succsim_\varphi}(B)$.

Summing up, we have shown $\varphi(B) = \varphi_{\succsim_\varphi}(B)$. □

2.2 Rationalizability by Maximization of a Linear Order

Now let us consider choice data without ties. Here choice data takes the form of a function $\phi : \mathcal{B} \to X$, which maps each opportunity set $B \in \mathcal{B}$ into one element $\phi(B) \in B$.

Since there are no ties, the natural preference ranking is a linear order which excludes indifference. A linear order is such that:

Trichotomy: For all $x, y \in X$, only one of $x \succ y$ or $y \succ x$ or $x = y$ holds.
Transitivity: For all $x, y, z \in X$, if $x \succ y$ and $y \succ z$, then $x \succ z$.

Definition 2.2 Choice function ϕ is rationalizable if there is a linear order $\succ$ over X such that

$$\phi(B) \succ y$$

holds for all $B \in \mathcal{B}$ and $y \in B \setminus \{\phi(B)\}$.

We write the contraction property for choice functions as below. Intuition is the same.

Contraction: For all $B, C \in \mathcal{B}$, if $C \subset B$ and $\phi(B) \in C$, then $\phi(C) = \phi(B)$.

Now we can characterize rationalizability as the maximization of a linear order.

Theorem 2.2 *Assume that $\mathcal{B}$ consists of all the non-empty subsets of X. Then, ϕ satisfies Contraction if and only if it is rationalizable as the maximization of some linear order $\succ$.*

Proof Again, it suffices to show sufficiency of the contraction property. Suppose ϕ satisfies Contraction. Define $\succ_\phi$ by

$$x \succ_\phi y \iff x = \phi(\{x, y\}).$$

First we show $\succ_\phi$ satisfies Trichotomy and Transitivity. Pick any $x, y \in X$ with $x \neq y$, then only one of $x = \phi(\{x, y\})$ or $y \neq \phi(\{x, y\})$ must be true. Hence if $x \succ_\phi y$ it must be that $y \not\succ_\phi x$, and vice versa. Hence Trichotomy is met.

To show Transitivity, suppose $x \succ_\phi y$ and $y \succ_\phi z$. Then, by definition of $\succ_\phi$ we have $x = \phi(\{x, y\})$ and $y = \phi(\{y, z\})$. Now there are three cases.

(i) If $x = \phi(\{x, y, z\})$, then from Contraction it follows that $x = \phi(\{x, z\})$, hence $x \succ_\phi z$.
(ii) If $y = \phi(\{x, y, z\})$, then from Contraction it follows that $y = \phi(\{x, y\})$, but this means $y \succ_\phi x$, which is a contradiction to $x \succ_\phi y$.
(iii) If $z = \phi(\{x, y, z\})$, then from Contraction it follows that $z = \phi(\{y, z\})$, but this means $z \succ_\phi y$, which contradicts $y \succ_\phi z$.

Finally, we show that ϕ is indeed rationalizable as the maximization of $\succ_\phi$. To see this, pick any $B \in \mathcal{B}$ and any $y \in B \setminus \{\phi(B)\}$, then from Contraction it follows that $\phi(B) = \phi(\{\phi(B), y\})$, which means $\phi(B) \succ_\phi y$. □

Chapter 3
Choice with Multiple Components and Separability

This chapter explains the nature of preferences when choice alternatives consist of multiple components. The central property is separability.

The set of alternatives X is given in the product form $X = \prod_{i \in I} X_i$, where I is the index set of components and X_i is the set of outcomes with component $i \in I$.

Example 3.1 The index set $I \subset \mathbb{N}$ refers to the set of time periods, and for each $i \in I$ the set X_i is interpreted as the set of outcomes in Period i.

We think in this way, because even if goods are materially the same when they are consumed at different time periods they are regarded as economically different goods. For example, one gallon of gasoline being consumed today and one gallon of gasoline being consumed one month later are economically different goods. Thus saving is understood as an exchange activity of buying future consumption by means of selling current consumption, and borrowing is understood as an exchange activity of buying current consumption by means of selling future consumption.

Example 3.2 Think of a situation under uncertainty and consider that the decision maker has to make some decision before resolution of uncertainty.

Then the index set I is seen as the set of possible states of the world, and for each $i \in I$ the set X_i is seen as the set of outcomes that can be given at state i.

We think in this way, because even if goods are materially the same when they are received and consumed at different states they are regarded as economically different goods. For example, one gallon of gasoline to be consumed in the next summer if it is hot and one gallon of gasoline to be consumed in the next summer if it is cold are different goods, and we exchange between them directly or indirectly through commodity markets and security markets.

Hereafter, assume that the index set I is finite, and that for each $i \in I$ the set X_i is a connected subset of a separable topological space, and X is given the product topology.

The following three axioms are standard.

T. Hayashi, *Decision Theory*, Monographs in Mathematical Economics 9,
https://doi.org/10.1007/978-981-95-2200-2_3

Completeness: For all $x, y \in X$, either $x \succsim y$ or $y \succsim x$.
Transitivity: For all $x, y, z \in X$, if $x \succsim y$ and $y \succsim z$ then $x \succsim z$.
Continuity: $\{(x, y) \in X \times X : x \succsim y\}$ is a closed subset of $X \times X$.

We introduce two kinds of separability conditions. For index $j \in I$, let $X_{-j} = \prod_{i \in I \setminus \{j\}} X_i$ and denote its element by $x_{-j} = (x_i)_{i \in I \setminus \{j\}}$.

Weak Separability: For all $j \in I$, for all $x_j, y_j \in X_j$ and $z_{-j}, w_{-j} \in X_{-j}$,

$$(x_j, z_{-j}) \succsim (y_j, z_{-j})$$
$$\text{if and only if } (x_j, w_{-j}) \succsim (y_j, w_{-j}).$$

The condition states that the preference ranking over alternatives in each specific component is determined independently of what are given regarding the other components.

The name "weak" separability suggests that there is a "strong" one. The strong one says that for any **subset** or group of components the preference ranking over alternatives regarding the set is determined independently of what are given regarding the complement.

For a subset of indices $J \subset I$, let $X_J = \prod_{i \in J} X_i$ and denote its element by $x_J = (x_i)_{i \in J} \in X_J$. Also, for the complement set $I \setminus J$, let $X_{-J} = \prod_{i \in I \setminus J} X_i$ and denote its element by $x_{-J} = (x_i)_{i \in I \setminus J}$.

Strong Separability: For all $J \subset I$, for all $x_J, y_J \in X_J$ and $z_{-J}, w_{-J} \in X_{-J}$,

$$(x_J, z_{-J}) \succsim (y_J, z_{-J})$$
$$\text{if and only if } (x_J, w_{-J}) \succsim (y_J, w_{-J}).$$

Weak separability is known to imply the following.

Theorem 3.1 *For preference $\succsim$ defined over X, the following two statements are equivalent.*

1. *$\succsim$ satisfies Completeness, Transitivity, Continuity and Weak Separability.*
2. *There exists a family of functions $\{u_i\}_{i \in I}$ with $u_i : X_i \to \mathbb{R}$ being continuous for each $i \in I$ and an increasing and continuous function $f : \prod_{i \in I} u_i(X_i) \to \mathbb{R}$ such that $\succsim$ is represented by a function $U : X \to \mathbb{R}$ in the form*

$$U(x) = f\left((u_i(x_i))_{i \in I}\right).$$

Proof Necessity of Completeness, Transitivity and Continuity are straightforward. Necessity of Weak Separability follows from

$$(x_i, z_{-i}) \succsim (y_i, z_{-i})$$
$$\iff f(u_i(x_i), (u_j(z_j))_{j \neq i}) \geq f(u_i(y_i), (u_j(z_j))_{j \neq i})$$

$$\begin{aligned}&\iff u_i(x_i) \geq u_i(y_i)\\&\iff f(u_i(x_i), (u_j(z'_j))_{j\neq i}) \geq f(u_i(y_i), (u_j(z'_j))_{j\neq i})\\&\iff (x_i, z'_{-i}) \succsim (y_i, z'_{-i})\end{aligned}$$

because f is increasing.

Now we prove sufficiency. For each $i \in I$, define the preference $\succsim_i$ induced over X_i by

$$x_i \succsim_i y_i \iff (x_i, z_{-i}) \succsim (y_i, z_{-i})$$

for some arbitrarily fixed $z_{-i} \in X_{-i}$. Because of Weak Separability, the definition of $\succsim_i$ does not depend on the choice of z_{-i}.

Since $\succsim$ and $(\succsim_i)_{i\in I}$ satisfy Completeness, Transitivity and Continuity, they have continuous representations U and $(u_i)_{i\in I}$, respectively. Fix them throughout.

Define a function $f : \prod_{i\in I} u_i(X_i) \to \mathbb{R}$ by

$$f((v_i)_{i\in I}) = U(x)$$

for each $(v_i)_{i\in I} \in \prod_{i\in I} u_i(X_i)$, where $x \in X$ is such that $u_i(x_i) = v_i$ holds for each $i \in I$.

Since any $x, y \in X$ such that $u_i(x_i) = u_i(y_i) = v_i$ for each $i \in I$ must yield $x_i \sim_i y_i$ for every $i \in I$, by Weak Separability we must have $x \sim y$. Hence the definition of f does not depend on the choice of x.

Also, since U and $(u_i)_{i\in I}$ are continuous, f is continuous too. To see that f is increasing, consider arbitrary $v_i, v'_i \in u_i(X_i)$ with $v_i > v'_i$ and $(v_j)_{j\neq i} \in \prod_{j\neq i} u_j(X_j)$. Then, by taking $x_i, x'_i \in X_i$ and $z_{-i} \in \prod_{j\neq i} X_j$ such that $u_i(x_i) = v_i$, $u_i(x'_i) = v'_i$ and $(u_j(z_j))_{j\neq i} = (v_j)_{j\neq i}$, we have $x_i \succ_i y_i$. Because of the definition of $\succsim_i$, we have $(x_i, z_{-i}) \succ_i (x'_i, z_{-i})$. By the definition of f, $f(u_i(x_i), (u_j(z_j))_{j\neq i}) > f(u_i(x'_i), (u_j(z_j))_{j\neq i})$, hence we have $f(v_i, (v_j)_{j\neq i}) > f(v'_i, (v_j)_{j\neq i})$. □

Strong separability is known to imply the following. From here we assume $|I| \geq 3$, since it is trivial when $|I| = 1$ and Strong Separability and Weak Separability are equivalent when $|I| = 2$.

Theorem 3.2 *Assume $|I| \geq 3$. Then, for preference $\succsim$ defined over X, the following two statements are equivalent:*

1. *$\succsim$ satisfies Completeness, Transitivity, Continuity and Strong Separability.*
2. *There exists a family of functions $\{u_i\}_{i\in I}$ with $u_i : X_i \to \mathbb{R}$ being continuous for each $i \in I$ such that $\succsim$ is represented by a function $U : X \to \mathbb{R}$ in the form*

$$U(x) = \sum_{i\in I} u_i(x_i).$$

Moreover, two family of functions $\{u_i\}_{i\in I}$ *and* $\{v_i\}_{i\in I}$ *represent the same preference in the above form if and only if there exist constants* $A > 0$ *and* $\{B_i\}_{i\in I}$ *such that*

$$v_i = Au_i + B_i$$

holds for every $i \in I$.

Remark 3.1 Note that the *entire* representation is still ordinal, and for any monotone function $f : \sum_{i\in I} u_i(X_i) \to \mathbb{R}$ the function given by

$$f(U(x)) = f\left(\sum_{i\in I} u_i(x_i)\right)$$

represents the same preference as $U(x) = \sum_{i\in I} u_i(x_i)$ does.

Proof **Existence of Representation** Since necessity of the axioms is straightforward, we prove sufficiency. The first and direct proof by Debreu is very hard to follow (you should try, though). Koopmans provides an elementary proof, but it is very long and requires lots of drawings. A more comprehensive treatment by Gorman requires lots of preliminaries which are beyond our scope. Therefore, instead, we limit ourselves to be content with borrowing a theorem on functional equation by von Stengel .

When $|I| = 3$ Define preference $\succsim_1$ induced over X_1 by

$$x_1 \succsim_1 y_1 \iff (x_1, z_2, z_3) \succsim (y_1, z_2, z_3),$$

where we fix some $(z_2, z_3) \in X_2 \times X_3$ arbitrarily. By Strong Separability, the definition of $\succsim_1$ does not depend on the choice of (z_2, z_3). Similarly for $\succsim_2, \succsim_3$.

Define preference $\succsim_{12}$ induced over $X_1 \times X_2$ by

$$(x_1, x_2) \succsim_{12} (y_1, y_2) \iff (x_1, x_2, z_3) \succsim (y_1, y_2, z_3),$$

where we fix some $z_3 \in X_3$ arbitrarily. By Strong Separability, the definition of $\succsim_{12}$ does not depend on the choice of z_3. Similarly for $\succsim_{13}, \succsim_{23}$.

Since all of $\succsim, \succsim_1, \succsim_2, \succsim_3, \succsim_{12}, \succsim_{13}, \succsim_{23}$ satisfy Completeness, Transitivity and Continuity, they have continuous representations $\tilde{U}, u_1, u_2, u_3, u_{12}, u_{13}, u_{23}$, respectively. We fix them throughout.

By applying Theorem 3.1 to separation of $\succsim$ to $\succsim_{12}$ and $\succsim_3$, there is an increasing and continuous function $F : u_{12}(X_1 \times X_2) \times u_3(X_3) \to \mathbb{R}$ such that

$$\tilde{U}(x_1, x_2, x_3) = F(u_{12}(x_1, x_2), u_3(x_3))$$

holds.

Also, by applying Theorem 3.1 to separation of $\succsim_{12}$ to $\succsim_1$ and $\succsim_2$, there is an increasing and continuous function $G : u_1(X_1) \times u_2(X_2) \to \mathbb{R}$ such that

$$u_{12}(x_1, x_2) = G(u_1(x_1), u_2(x_2)),$$

holds.

Summing up, we obtain

$$\widetilde{U}(x_1, x_2, x_3) = F(G(u_1(x_1), u_2(x_2)), u_3(x_3)).$$

Likewise, by applying Theorem 3.1 to separation of $\succsim$ to $\succsim_1$ and $\succsim_{23}$, there is an increasing and continuous function $H : u_1(X_1) \times u_{23}(X_2 \times X_3) \to \mathbb{R}$ such that

$$\widetilde{U}(x_1, x_2, x_3) = H(u_1(x_1), u_{23}(x_2, x_3))$$

holds.

Also, by applying Theorem 3.1 to separation of $\succsim_{23}$ to $\succsim_2$ and $\succsim_3$, there is an increasing and continuous function $K : u_2(X_2) \times u_3(X_3) \to \mathbb{R}$ such that

$$u_{23}(x_2, x_3) = K(u_2(x_2), u_3(x_3)).$$

Summing up, we obtain

$$\widetilde{U}(x_1, x_2, x_3) = H(u_1(x_1), K(u_2(x_2), u_3(x_3))).$$

Denote the ranges of u_1, u_2, u_3 by U_1, U_2, U_3, respectively. Then, by connectedness of X_1, X_2, X_3 and continuity of u_1, u_2, u_3, they are connected intervals.

Therefore, for arbitrary $(u_1, u_2, u_3) \in U_1 \times U_2 \times U_3$,

$$F(G(u_1, u_2), u_3) = H(u_1, K(u_2, u_3)).$$

From von Stengel , the solution of the above functional equation has the following form: There exist functions $\phi_1 : U_1 \to \mathbb{R}$, $\phi_2 : U_2 \to \mathbb{R}$, $\phi_3 : U_3 \to \mathbb{R}$, $\phi_{12} : \phi_1(U_1) + \phi_2(U_2) \to \mathbb{R}$, $\phi_{23} : \phi_2(U_2) + \phi_3(U_3) \to \mathbb{R}$, $\phi : \phi_1(U_1) + \phi_2(U_2) + \phi_3(U_3) \to \mathbb{R}$ such that

$$\begin{aligned}
F(u_{12}, u_3) &= \phi(\phi_{12}^{-1}(u_{12}) + \phi_3(u_3)), \\
G(u_1, u_2) &= \phi_{12}(\phi_1(u_1) + \phi_2(u_2)), \\
H(u_1, u_{23}) &= \phi(\phi_1(u_1) + \phi_{23}^{-1}(u_{23})), \\
K(u_2, u_3) &= \phi_{23}(\phi_2(u_2) + \phi_3(u_3))
\end{aligned}$$

holds.

Now, because

$$\phi(\phi_{12}^{-1}(\phi_{12}(\phi_1(u_1) + \phi_2(u_2))) + \phi_3(u_3)) = \phi(\phi_1(u_1) + \phi_2(u_2) + \phi_3(u_3)),$$

for every $(x_1, x_2, x_3) \in X_1 \times X_2 \times X_3$ we obtain

$$\widetilde{U}(x_1, x_2, x_3) = \phi(u_1(x_1) + u_2(x_2) + u_3(x_3)).$$

Finally, define $U : X_1 \times X_2 \times X_3 \to \mathbb{R}$ by

$$U(x_1, x_2, x_3) = \phi^{-1}(\widetilde{U}(x_1, x_2, x_3)).$$

then it is the desired representation.

When $|I| \geq 3$ We prove by induction. Let $I = \{1, 2, \cdots, k\}$, and assume that the conclusion is correct up to $k = n - 1$, and now let $k = n$.

Apply the result for the case of $|I| = 3$ to separation of $I = \{1, 2, \cdots, n\}$ to a triple $\{1, \cdots, n-2\}$, $\{n-1\}$, $\{n\}$, then we obtain additive representation

$$u_{\{1,\cdots,n-2\}}(x_1, \cdots, x_{n-2}) + u_{n-1}(x_{n-1}) + u_n(x_n).$$

Also, define $\precsim_{\{1,\cdots,n-1\}}$ induced over $\prod_{i \in \{1,\cdots,n-1\}} X_i$ by

$$\begin{aligned}&(x_1, \cdots, x_{n-1}) \precsim_{\{1,\cdots,n-1\}} (y_1, \cdots, y_{n-1})\\ \Longleftrightarrow\ &(x_1, \cdots, x_{n-1}, x_n) \precsim (y_1, \cdots, y_{n-1}, x_n),\end{aligned}$$

where we fix some $x_n \in X_n$ arbitrarily. Then, by Strong Separability this definition does not depend on the choice of x_n. Then, since $\precsim_{\{1,\cdots,n-1\}}$ satisfies Strong Separability over $\prod_{i \in \{1,\cdots,n-1\}} X_i$, by the induction assumption it has additive representation

$$u_1^*(x_1) + \cdots + u_{n-2}^*(x_{n-2}) + u_{n-1}^*(x_{n-1}).$$

Now, since

$$u_{\{1,\cdots,n-2\}}(x_1, \cdots, x_{n-2}) + u_{n-1}(x_{n-1})$$

and

$$u_1^*(x_1) + \cdots + u_{n-2}^*(x_{n-2}) + u_{n-1}^*(x_{n-1})$$

represent the same $\succsim_{\{1,\cdots,n-1\}}$ in additive form, from the uniqueness result shown below, there exist constants $A > 0$ and B, B' such that

$$u_{\{1,\cdots,n-2\}}(x_1,\cdots,x_{n-2}) = A(u_1^*(x_1) + \cdots + u_{n-2}^*(x_{n-2})) + B$$
$$u_{n-1}(x_{n-1}) = Au_{n-1}^*(x_{n-1}) + B'.$$

Hence, by taking

$$u_i(x_i) = Au_i^*(x_i) + B_i$$

for each $i \in \{1,\cdots,n-2\}$ so that $\sum_{i\in\{1,\cdots,n-2\}} B_i = B$, the representation of $\succsim$ becomes

$$u_1(x_1) + \cdots + u_{n-2}(x_{n-2}) + u_{n-1}(x_{n-1}) + u_n(x_n).$$

Uniqueness Assume $|I| = 3$ again, since the general case is analogous. Consider two additive representations of the same preference,

$$U(x_1, x_2, x_3) = u_1(x_1) + u_2(x_2) + u_3(x_3),$$
$$U^*(x_1, x_2, x_3) = u_1^*(x_1) + u_2^*(x_2) + u_3^*(x_3).$$

Fix some $(\widehat{x}_1, \widehat{x}_2, \widehat{x}_3) \in X_1 \times X_2 \times X_3$ arbitrarily, and let

$$B_1 = u_1^*(\widehat{x}_1),\quad B_2 = u_2^*(\widehat{x}_2),\quad B_3 = u_3^*(\widehat{x}_3).$$

Define

$$u_1^{**}(x_1) = u_1^*(x_1) - B_1,\quad u_2^{**}(x_2) = u_2^*(x_2) - B_2,\quad u_3^{**}(x_3) = u_3^*(x_3) - B_3,$$

then it is also an additive representation of the same preference, there is an increasing and continuous function h such that

$$u_1^{**}(x_1) + u_2^{**}(x_2) + u_3^{**}(x_3) = h(u_1(x_1) + u_2(x_2) + u_3(x_3))$$

holds for any x_1, x_2, x_3.

Fix $x_2 = \widehat{x}_2$, $x_3 = \widehat{x}_3$, then

$$u_1^{**}(x_1) = h(u_1(x_1))$$

for all x_1. Likewise,

$$u_2^{**}(x_2) = h(u_2(x_2))$$

for all x_2.

Also, fix $x_3 = \widehat{x}_3$, then

$$u_1^{**}(x_1) + u_2^{**}(x_2) = h(u_1(x_1) + u_2(x_2))$$

for all x_1, x_2.

Therefore, for any $u_1 \in U_1, u_2 \in U_2$ we have

$$h(u_1 + u_2) = h(u_1) + h(u_2),$$

which is Cauchy's functional equation. The only solution is that h is linear, that is, there is $A > 0$ such that

$$h(u) = Au.$$

□

Chapter 4
Decision Under Risk

The distinction between risk and uncertainty seems standard nowadays, but let me briefly note here that risk refers to situations in which probability distribution is given objectively, as in coin flipping or throwing a die. On the other hand, uncertainty refers to situations in which no such thing is given as an object.

This chapter introduces expected utility theory, the standard theory of decision making under risk.

4.1 Risk Aversion and Expected Utility Theory

The first thing we think of would be to compare expected values of outcomes, but this leads to problems like below.

Example 4.1 If we think of expected values of receipts, then the decision maker will be indifferent for example between receiving 50 dollars for sure and receiving 100 dollars with probability 0.5. Or if the sure receipt is 49.99 dollars he would take 100 dollars definitely. But this is unrealistic.

Example 4.2 (St. Petersburg Paradox) Flip a coin repeatedly as long as you have heads. If you have heads k times then you receive 2^k dollars. Then, the probability that you receive 2^k is the probability that you have heads k times and then tail which is 2^{k+1}. Hence the expected value of money you receive is

$$\sum_{k=0}^{\infty} \frac{2^k}{2^{k+1}} = \sum_{k=0}^{\infty} \frac{1}{2} = \infty.$$

Now, are you willing to pay let's say million dollars to attend this game?

T. Hayashi, *Decision Theory*, Monographs in Mathematical Economics 9,
https://doi.org/10.1007/978-981-95-2200-2_4

The above two examples suggest that it is realistic and also convincing to take higher order properties of probability distributions, such as variance, not just expected value. We relegate the precise definition to a later section, but you are said to be *risk-averse* if you prefer a sure receipt when expected values are the same, you are said to be *risk-loving* if you rather prefer to bet, and *risk-neutral* if you are indifferent and consider only expected values.

The expected utility theory was established in order to systematically and quantitatively describe such risk attitudes.

Let Z denote the set of outcomes. It could be any set, but in applications we mostly consider that $Z = \mathbb{R}$ or $Z = \mathbb{R}_+$.

Then, a probability distribution l over Z is said to be a *simple lottery* if it assigns positive probabilities to only finite number of outcomes. That is, for a simple lottery l, there is a finite set $S(l) \subset Z$ such that it assigns $l(z) > 0$ to each $z \in S(l)$ and $\sum_{z \in S(l)} l(z) = 1$ holds. Denote the set of simple lotteries over Z by $\Delta_S(Z)$.[1]

Denote a preference relation over $\Delta_S(Z)$ by $\succsim$.

Definition 4.1 Preference $\succsim$ is said to have an expected utility representation if there is a function $u : Z \to \mathbb{R}$ such that

$$l \succsim m \iff \sum_{z \in S(l)} u(z)l(z) \geqq \sum_{z \in S(m)} u(z)m(z)$$

holds for all $l, m \in \Delta_S(Z)$.

We call the function u a von Neumann/Morgenstern index (vNM index).

Let us suppose $Z = \mathbb{R}_+$ and consider the case of monetary receipts. If u is affine, that is, $u(z) = az + b$, then

$$\sum_{z \in S(l)} u(z)l(z) = \sum_{z \in S(l)} (az + b)l(z) = a \sum_{z \in S(l)} zl(z) + b$$

and the decision maker cares only for the expected value of receipt $\sum_{z \in S(l)} zl(z)$. Thus he is risk-neutral.

On the other hand, suppose u is strictly concave, that is, for every two distinct $z, z' \in \mathbb{R}_+$ and for all $\lambda \in (0, 1)$,

$$u(\lambda z + (1 - \lambda)z') > \lambda u(z) + (1 - \lambda)u(z').$$

Under the expected utility theory, this means receiving $\lambda z + (1 - \lambda)z'$ for sure is better than receiving z with probability λ and z' with probability $1 - \lambda$. That is, the decision maker is risk-averse.[2]

---

[1] The lottery as in the St. Petersburg paradox is not simple, and therefore non-simple lotteries will be explained in a later section.

[2] If u is weakly concave, that is, for all $z, z' \in \mathbb{R}_+$ and $\lambda \in [0, 1]$,

Also, suppose u is strictly convex, that is, for every two distinct $z, z' \in \mathbb{R}_+$ and for all $\lambda \in (0, 1)$,

$$u(\lambda z + (1-\lambda)z') < \lambda u(z) + (1-\lambda)u(z').$$

Under the expected utility theory, this means receiving z with probability λ and z' with probability $1-\lambda$ is better than receiving $\lambda z + (1-\lambda)z'$ for sure. That is, the decision maker is risk-loving.

4.2 Axiomatic Characterization of Expected Utility Theory

We continue to consider the set of simple lotteries over the set of outcomes Z, that is, $\Delta_S(Z)$. For any two simple lotteries $l, m \in \Delta_S(Z)$ and a number $\alpha \in [0, 1]$, define the **mixture** $\alpha l + (1-\alpha)m$ as the simple lottery given by

$$(\alpha l + (1-\alpha)m)(z) = \alpha l(z) + (1-\alpha)m(z)$$

for each $z \in S(l) \cup S(m)$.

We continue to consider the preference relation over $\Delta_S(Z)$, which is denoted by $\succsim$, and consider the following axioms:

Completeness: For all $l, m \in \Delta_S(Z)$, either $l \succsim m$ or $m \succsim l$.
Transitivity: For all $l, m, n \in \Delta_S(Z)$, if $l \succsim m$ and $m \succsim n$, then $l \succsim n$.
Mixture Continuity: For all $l, m, n \in \Delta_S(Z)$, the sets $\{\lambda \in [0, 1] : \lambda l + (1-\lambda)m \succsim n\}$ and $\{\lambda \in [0, 1] : n \succsim \lambda l + (1-\lambda)m\}$ are closed subsets of $[0, 1]$.
Independence: For all $l, m, n \in \Delta_S(Z)$ and for all $\lambda \in (0, 1)$,

$$l \succsim m \text{ if and only if } \lambda l + (1-\lambda)n \succsim \lambda m + (1-\lambda)n.$$

An "intuitive" explanation of the independence axiom is:

> If the decision maker prefers lottery l over lottery m, then he would prefer a "lottery" that gives l with probability α and n with probability $1-\alpha$ over a "lottery" that gives m with probability α and n with probability $1-\alpha$, and the converse is true as well.

Otherwise, suppose let's say the decision maker prefers the "lottery" that gives m with probability α and n with probability $1-\alpha$ over the "lottery" that gives l with probability α and n with probability $1-\alpha$, and if the resolution of those "lotteries" is that m and l are obtained respectively, he would change his mind and now wants l, which is a dynamic inconsistency.

$$u(\lambda z + (1-\lambda)z') \geqq \lambda u(z) + (1-\lambda)u(z'),$$

then we can characterize weak risk aversion which includes risk neutrality as a special case.

Note, however, that this interpretation implicitly presumes that the "lottery" that gives l with probability α and n with probability $1-\alpha$ and the mixture $\alpha l+(1-\alpha)n$ are the same. We come back to this issue later.

4.3 Expected Utility Representation Theorem

We follow the classic argument by Herstein and Milnor [62]. We first show several lemmata.

Lemma 4.1 *Suppose $a \succ b$, then:*

(i) *for all $0 < \lambda < 1$, $a \succ \lambda a + (1-\lambda)b \succ b$, and*
(ii) *$\lambda a + (1-\lambda)b \succ \eta a + (1-\eta)b \iff \lambda > \eta$.*

Proof

(i) By Independence, we obtain $a = \lambda a + (1-\lambda)a \succ \lambda a + (1-\lambda)b$ and $\lambda a + (1-\lambda)b \succ \lambda b + (1-\lambda)b = b$.
(ii) When $\lambda > \eta$, it must be that $\eta < 1$. Hence from (i) we have $a \succ \eta a+(1-\eta)b \succ b$. Now by applying (i) to $a \succ \eta a + (1-\eta)b$, we obtain

$$\begin{aligned}\lambda a + (1-\lambda)b &= \frac{\lambda-\eta}{1-\eta}a + \frac{1-\lambda}{1-\eta}(\eta l + (1-\eta)b)\\ &\succ \eta a + (1-\eta)b.\end{aligned}$$

The converse is similar.

□

Lemma 4.2 *If $a \succsim c \succsim b$ and $a \succ b$, there is a unique $0 \le \lambda \le 1$ such that $\lambda a + (1-\lambda)b \sim c$.*

Proof Suppose there is no such λ. Then the closed interval $[0, 1]$ is partitioned into $\{\lambda \in [0, 1] : \lambda a + (1-\lambda)b \succ c\}$ and $\{\lambda \in [0, 1] : c \succ \lambda a + (1-\lambda)b\}$, which are open subsets of $[0, 1]$ with respect to the relative topology. Hence it contradicts the connectedness of $[0, 1]$.

Uniqueness follows from the previous lemma. □

Lemma 4.3 *If $a \succsim b$ and $c \succsim d$, then $\lambda a + (1-\lambda)c \succsim \lambda b + (1-\lambda)d$ for all $\lambda \in [0, 1]$.*

Proof By Independence we have $\lambda a+(1-\lambda)c \succsim \lambda a+(1-\lambda)d$ and $\lambda a+(1-\lambda)d \succsim \lambda b + (1-\lambda)d$. By Transitivity the conclusion follows. □

Lemma 4.4 *Let $\succsim$ be a binary relation defined over $\Delta_S(Z)$. Then the following two statements are equivalent:*

(a) *$\succsim$ satisfies Completeness, Transitivity, Mixture Continuity and Independence.*

(b) $\succsim$ *is represented by some mixture-linear function* $U : \Delta_S(Z) \to \mathbb{R}$*. That is, for all* $a, b \in \Delta_S(Z)$ *and* $\lambda \in [0, 1]$*,*

$$U(\lambda a + (1-\lambda)b) = \lambda U(a) + (1-\lambda)U(b).$$

If V *is also a mixture-linear representation of* $\succsim$*, then there exist* $\alpha > 0$ *and* β *such that*

$$V = \alpha U + \beta.$$

Proof Necessity of each axiom is obvious and we show sufficiency of the four axioms.

The statement is trivial when the preference is completely indifferent, hence suppose there exist $\overline{a}, \underline{a}$ such that $\overline{a} \succ \underline{a}$.

Set $U(\overline{a})$ and $U(\underline{a})$ arbitrarily. For simplicity, let $U(\overline{a}) = 1$ and $U(\underline{a}) = 0$. Then for every $a \in \Delta_S(Z)$, we consider the following three cases:

Case 1: If $\overline{a} \succsim a \succsim \underline{a}$, then define $U(l) = \lambda$, where λ is such that $\lambda\overline{a} + (1-\alpha)\underline{a} \sim a$ holds. By Lemma 4.2, such a λ is uniquely determined.

Case 2: If $a \succ \overline{a}$, then define $U(a) = 1/\lambda$, where λ is such that $\lambda a + (1-\lambda)\underline{a} \sim \overline{a}$ holds. By Lemma 4.2, such a λ is uniquely determined.

Case 3: If $\underline{a} \succ a$, then define $U(a) = 1/(1 - 1/\lambda)$, where λ is such that $\lambda\overline{a} + (1-\lambda)a \sim \underline{a}$ holds. By Lemma 4.2, such a λ is uniquely determined.

To show mixture-linearity of U, consider let's say the case that $a \sim U(a)\overline{a} + (1-U(a))\underline{a}$ and $b \sim U(b)\overline{a} + (1-U(b))\underline{a}$. Then from Lemma 4.3 it follows that

$$\begin{aligned}
&\lambda a + (1-\lambda)b \\
&\sim \lambda\left(U(a)\overline{a} + (1-U(a))\underline{a}\right) + (1-\lambda)\left(U(b)\overline{a} + (1-U(b))\underline{a}\right) \\
&\sim (\lambda U(a) + (1-\lambda)U(b))\overline{a} + \{1 - (\lambda U(a) + (1-\lambda)U(b))\}\,\underline{a}.
\end{aligned}$$

Similarly for the other cases.

Uniqueness Up to Positive Affine Transformations Suppose V is also a mixture-linear representation of $\succsim$.

Then, let $\alpha = V(\overline{a}) - V(\underline{a})$ and $\beta = V(\underline{a})$. By mixture-linearity, if $a \sim \lambda\overline{a} + (1-\lambda)\underline{a}$ let's say then that

$$V(a) = \lambda V(\overline{a}) + (1-\alpha)V(\underline{a}) = U(a)V(\overline{a}) + (1-U(a))V(\underline{a}) = \alpha U(a) + \beta$$

Similarly for the other cases. □

Now we present the expected utility representation theorem.

Theorem 4.1 *For a preference relation* $\succsim$ *defined over the set of simple lotteries* $\Delta_S(Z)$*, the following two statements are equivalent:*

1. $\succsim$ *satisfies Completeness, Transitivity, Mixture Continuity and Independence.*
2. *There exists a function* $u : Z \to \mathbb{R}$ *such that* $\succsim$ *is represented by* $U : \Delta_S(Z) \to \mathbb{R}$ *in the form*

$$U(l) = \sum_{z \in S(l)} u(z) l(z).$$

Moreover, two functions u, v *give representations in the above form if and only if there is are numbers* A, B *with* $A > 0$ *and*

$$v = Au + B.$$

Proof Necessity of the axioms is obvious. We show sufficiency.

From Lemma 4.4, $\succsim$ has a mixture-linear representation $U : \Delta_S(Z) \to \mathbb{R}$.

For all $z \in Z$, $\delta(z)$ denote the lottery that gives z with probability 1. Then, define $u : Z \to \mathbb{R}$ by $u(z) = U(\delta(z))$. Then for any lottery with two outcomes, $l = (x_1; l_1, x_2; l_2)$, by mixture-linearity we have

$$U(l) = l_1 U(\delta(x_1)) + l_2 U(\delta(x_2)) = l_1 u(x_1) + l_2 u(x_2).$$

For any lottery with three outcomes, $l = (x_1; l_1, x_2; l_2, x_3; l_3)$, again by mixture-linearity and the result obtained just now we have

$$\begin{aligned} U(l) &= l_1 U(\delta(x_1)) + (l_2 + l_3) U\left(x_2; \frac{l_2}{l_2 + l_3}, x_3; \frac{l_3}{l_2 + l_3}\right) \\ &= l_1 U(\delta(x_1)) + (l_2 + l_3)\left(\frac{l_2}{l_2 + l_3} U(\delta(x_2)) + \frac{l_3}{l_2 + l_3} U(\delta(x_3))\right) \\ &= l_1 U(\delta(x_1)) + l_2 U(\delta(x_2)) + l_3 U(\delta(x_3)) \\ &= l_1 u(x_1) + l_2 u(x_2) + l_3 u(x_3). \end{aligned}$$

For any $l \in \Delta_S(Z)$, since it is given in the form $l = \sum_{z \in S(l)} l(z)\delta(z)$ we can obtain the result by repeatedly applying the above argument.

Uniqueness follows from the previous result. □

4.4 Expected Utility over Non-simple Lotteries*

We need to think of non-simple lotteries typically when outcomes are continuous. Let us extend the expected utility theory to the domain of non-simple lotteries, following Grandmont [50].

Let (Z, d) be a compact metric space with respect to distance d, and let Σ denote the Borel σ-field generated by d. Then let $\Delta(Z)$ denote the set of Borel probability measures over (Z, Σ).

Let $B(z, \varepsilon) = \{y \in Z : d(y, z) < \varepsilon\}$ denote the open ball around $z \in Z$ with radius $\varepsilon > 0$, and let

$$Y^{\varepsilon} = \bigcup_{z \in Y} B(z, \varepsilon)$$

denote the ε-neighborhood of $Y \subset Z$.

We consider weak convergence as the convergence criterion. Say that a sequence of Borel probability measures $\{l^{\nu}\}$ converges weakly to a Borel probability measure l if

$$\int_Z f(x) dl^{\nu}(x) \to \int_Z f(x) dl(x)$$

holds for all bounded and continuous functions $f : Z \to \mathbb{R}$, while boundedness follows from continuity since Z is assumed to be compact here.

Weak convergence of the Borel probability measure is known to be equivalent to convergence in the so-called Prokhorov metric

$$\pi(l, m) = \inf \left\{ \varepsilon > 0 : l(Y) \leq m(Y^{\varepsilon}) + \varepsilon,\ m(Y) \leq l(Y^{\varepsilon}) + \varepsilon, \forall Y \in \Sigma \right\}$$

and $\Delta(Z)$ is known to be compact in the Prokhorov metric when Z is compact.

Under the weak convergence topology, the set of simple lotteries is known to be dense in $\Delta(X)$ in the following sense.

Proposition 4.1 *For all $l \in \Delta(Z)$ there is a sequence of simple lotteries $\{l^{\nu}\}$ such that*

$$\int_Z f(x) dl^{\nu}(x) \to \int_Z f(x) dl(x)$$

holds for any bounded and continuous function $f : Z \to \mathbb{R}$.

We impose the following axioms, which are the same as before except for replacing Mixture Continuity with Continuity, a stronger one.

Completeness: For all $l, m \in \Delta(Z)$, either $l \succsim m$ or $m \succsim l$.
Transitivity: For all $l, m, n \in \Delta(Z)$, if $l \succsim m$ and $m \succsim n$ then $l \succsim n$.

Continuity: The set $\{(l, m) \in \Delta(Z) \times \Delta(Z) : l \succsim m\}$ is a closed subset of $\Delta(Z) \times \Delta(Z)$.

Independence: For all $l, m, n \in \Delta(Z)$ and $\lambda \in (0, 1]$,

$$l \succsim m \text{ if and only if } \lambda l + (1 - \lambda)n \succsim \lambda m + (1 - \lambda)n.$$

Here is the expected utility representation theorem for non-simple lotteries.

Theorem 4.2 *For preference ranking $\succsim$ defined over $\Delta(Z)$, the following two statements are equivalent:*

1. *$\succsim$ satisfies Completeness, Transitivity, Continuity and Independence.*
2. *There exists a continuous function $u : Z \to \mathbb{R}$ such that $\succsim$ is represented by $U : \Delta(Z) \to \mathbb{R}$ in the form*

$$U(l) = \int_Z u(z)l(dz).$$

Moreover, if two functions u, v form representations in the above form if and only if there exist numbers A, B with $A > 0$ such that

$$v = Au + B$$

holds.

Proof Necessity of the axiom and uniqueness of representation are obvious, so we prove sufficiency here.

Since we are imposing Continuity, which is stronger than Mixture Continuity, we have the counterpart of Lemma 4.4 on the set of Borel probability measures. Hence there is a continuous and mixture-linear function $U : \Delta(Z) \to \mathbb{R}$ such that

$$l \succsim l' \iff U(l) \geq U(l')$$

holds.

Next, define $u : Z \to \mathbb{R}$ by

$$u(z) = U(\delta(z)).$$

Note that $\delta(z) \in \Delta(Z)$ refers to the probability measure that assigns probability 1 on z. Then, since it holds

$$\{y \in Z : u(y) \geq u(z)\} = \{y \in Z : \delta(y) \succsim \delta(z)\}$$

and

$$\{y \in Z : u(y) \leq u(z)\} = \{y \in Z : \delta(y) \precsim \delta(z)\}$$

by Continuity we obtain that u is a continuous function. Since Z is compact, u is bounded.

Now for simple lotteries we already know that

$$U(l) = \int_Z u(z)l(dz) = \sum_{z \in S(l)} u(z)l(z)$$

holds.

Pick an arbitrary non-simple lottery $l \in \Delta(Z)$. Then there exists a sequence of simple lotteries $\{l^\nu\}$ that converges to l in the weak convergence topology. Now by continuity, we obtain

$$\begin{aligned} U(l) &= \lim_{\nu \to \infty} U(l^\nu) \\ &= \lim_{\nu \to \infty} \int_Z u(z)l^\nu(dz) \\ &= \int_Z u(z)l(dz). \end{aligned}$$

□

4.5 Comparative Risk Aversion

In the expected utility theory, the curvature of the vNM index u in the representation

$$U(l) = \sum_{z \in S(l)} u(z)l(z)$$

describes the degree of risk aversion of the preference being represented.

Below is the definition of comparative risk aversion (Arrow [7] and Pratt [87]).

Definition 4.2 Say that $\succsim_1$ is more risk-averse than $\succsim_2$ if

$$l \succsim_1 \delta(x) \quad \Longrightarrow \quad l \succsim_2 \delta(x)$$

and

$$l \succ_1 \delta(x) \quad \Longrightarrow \quad l \succ_2 \delta(x)$$

hold for all $l \in \Delta(X)$ and $x \in X$.

That is, we define that 1 is more risk-averse than 2 by saying "once 1 takes a risk and chooses to bet it must be that 2 does so as well."

This definition of comparative risk aversion is meaningful only when those preferences coincide over deterministic outcomes. In this sense it is an incomplete ranking.

Lemma 4.5 *If* $\succsim_1$ *is more risk-averse than* $\succsim_2$*, then*

$$\delta(x) \succsim_1 \delta(y) \iff \delta(x) \succsim_2 \delta(y)$$

for all $x, y \in X$.

Proof By regarding $\delta(x)$ as the general lottery and $\delta(y)$ as the sure outcome, applying the first condition in the definition of comparative risk aversion, we obtain

$$\delta(x) \succsim_1 \delta(y) \implies \delta(x) \succsim_2 \delta(y).$$

By regarding $\delta(x)$ as the sure outcome and $\delta(y)$ as the general lottery, applying the second condition in the definition of comparative risk aversion, we obtain

$$\delta(y) \succ_1 \delta(x) \implies \delta(y) \succ_2 \delta(x),$$

that is,

$$\delta(x) \succsim_2 \delta(y) \implies \delta(x) \succsim_1 \delta(y).$$

□

Now being more risk-averse is characterized as having vNM index to be more concave.

Theorem 4.3 *Assume that* $\succsim_1$ *and* $\succsim_2$ *satisfy the expected utility theory, and that 1 is more risk-averse than 2. Let* $u_1, u_2 : X \to \mathbb{R}$ *be vNM indices which form expected utility representation of* $\succsim_1$ *and* $\succsim_2$*, respectively. Also, assume that for all* $x, y \in X$ *and* $\lambda \in [0, 1]$ *there exists* $z \in X$ *such that* $\delta(z) \sim_2 \lambda\delta(x) + (1 - \lambda)\delta(y)$.[3]

Then, there is an increasing and concave function $\varphi : u_2(X) \to u_1(X)$ *such that*

$$u_1(x) = \varphi(u_2(x)).$$

Proof By Lemma 4.5, u_1 and u_2 are ordinally equivalent. Hence there is a monotone function $\varphi : u_2(X) \to u_1(X)$ such that

$$u_1(x) = \varphi(u_2(x)).$$

Now, pick any $v, v' \in u_2(X)$ and $\lambda \in [0, 1]$, and pick $x, y, z \in X$ such that $u_2(x) = v$, $u_2(y) = v'$ and $u_2(z) = \lambda v + (1 - \lambda)v'$. Then, since $\lambda\delta(x) + (1 -$

[3] This assumption is met when X is connected and $\succsim_2$ is continuous.

$\lambda)\delta(y) \sim_2 \delta(z)$, from the second condition of the definition of comparative risk aversion we have $\lambda\delta(x) + (1-\lambda)\delta(y) \precsim_1 \delta(z)$. Hence it follows that

$$\begin{aligned}\varphi(\lambda v + (1-\lambda)v') &= \varphi(\lambda u_2(x) + (1-\lambda)u_2(y)) \\ &= \varphi(u_2(z)) \\ &= u_1(z) \\ &\geq \lambda u_1(x) + (1-\lambda)u_1(x) \\ &= \lambda\varphi(u_2(x)) + (1-\lambda)\varphi(u_2(y)) \\ &= \lambda\varphi(v) + (1-\lambda)\varphi(v').\end{aligned}$$

□

4.6 Is the Expected Utility "Cardinal?"

In the expected utility representation

$$U(l) = \sum_{z \in S(l)} u(z)l(z),$$

the curvature of the vNM index u explains the degree of risk aversion, and it is unique up to positive affine transformation, not with arbitrary monotone transformations.

For example, suppose $Z = \mathbb{R}_+$ and consider $u(z) = z$, that is, the preference represented by

$$U(l) = \sum_{z \in S(l)} zl(z)$$

is risk-neutral. However, if we take a monotone transformation of u by $\widetilde{u} = \sqrt{u}$, the preference represented by

$$\widetilde{U}(l) = \sum_{z \in S(l)} \widetilde{u}(z)l(z) = \sum_{z \in S(l)} \sqrt{z}l(z)$$

is now risk-averse. In this case, the function u is cardinal, in the sense that its curvature has a quantitative meaning.

Note, however, again that it has no meaning like a measure of happiness unless we bring in an extra assumption, since when u forms an expected utility representation $Au + B$ also forms an expected utility representation of the same risk preference.

Also, note that when we look at the *whole representation*, after taking any arbitrary monotone transformation f, the function

$$f(U(l)) = f\left(\sum_{z\in S(l)} u(z)l(z)\right)$$

represents the same preference over lotteries. Hence the whole representation is still ordinal.

For example, one could take an exponential transformation and obtain

$$V(l) = e^{U(l)} = e^{\sum_{z\in S(l)} u(z)l(z)} = \prod_{z\in S(l)} e^{u(z)p(z)} = \prod_{z\in S(l)} v(z)^{p(z)},$$

where $v(z) = u^{u(z)}$, which is a multiplicative form.

In other words, a preference satisfying the axioms of expected utility theory allows expected utility representations, but it does not exclude other forms of representation. And, we are saying that uniqueness up to positive affine transformation holds *when we restrict attention to expected utility representations of a preference which allows expected utility representations.* This is the key to the so-called Harsanyi-Sen debate on utilitarianism, which is nicely surveyed by Weymark [100].

4.7 Non-expected Utility Preference and Dynamic Consistency

Because the independence axiom is an "independence" axiom, it is hard to meet in reality. Let us look at the following example.

Example 4.3 (A Version of the Allais Paradox, Originally due to Allais [4])

(A): Choose from the two below:

A1: 30 dollars for sure
A2: 45 dollars with probability 0.8

(B): Choose from the two below:

B1: 30 dollars with probability 0.25
B2: 45 dollars with probability 0.2

An intuitive choice will be A1 in (A) and B2 in (B), and it is confirmed in experiments as well.

When we write such rankings formally, we have

$$(3000; 1) \succ (4500; 0.8, 0; 0.2),$$
$$(3000; 0.25, 0; 0.75) \prec (4500; 0.2, 0; 0.8).$$

But notice that the latter two objects are written as

$$(3000; 0.25, 0; 0.75) = 0.25(3000; 1) + 0.75(0; 1),$$
$$(4500; 0.2, 0; 0.8) = 0.25(4500; 0.8, 0; 0.2) + 0.75(0; 1).$$

Hence we see a violation of the independence axiom.

The expected utility theory weakens linearity of risk evaluation in outcomes (i.e., risk neutrality), but maintains linearity of risk evaluation in probabilities. Thus, the effect of increasing winning probability let's say from 0.4 to 0.5 is taken to be the same as the effect of increasing from 0.9 to 1.

The above example, however, shows that risk evaluation is not linear in probabilities and that the effect of winning probability is greater when it is closer to 1. This is called the *certainty effect* (Tversky and Kahneman [97]).

Also, there is a normative view that the independence axiom should be violated (Diamond [26]). Let's see the following example.

Example 4.4 There are two prisoners, A and B. You have to kill one of the two tomorrow. You are indifferent between killing either of the two prisoners.

Now suppose you think you should give them equality of chances. That is, you think flipping a coin and killing A with probability 0.5 and B with probability 0.5 is better than killing A with probability 1 arbitrarily and than killing B with probability 1 arbitrarily.

This contradicts the expected utility theory, however. For, under the independence axiom, if you are indifferent between killing A with probability 1 arbitrarily and killing B with probability 1 arbitrarily, then taking an even-chances mixture does not change the level of desirability.

This point was originally made in the context of criticizing Harsanyi's argument that a social welfare function should satisfy the expected utility theory (Harsanyi [55]). As far was a social welfare function is assumed to satisfy the expected utility theory, it does not admit the significance of "equality/fairness of chances" by means of lotteries.

You might say, "Why not simply drop the independence axiom?" However, we cannot just drop or weaken the independence axiom without causing a problem. Let us revisit the Allais paradox.

Example 4.5 (Revisiting the Allais Paradox)

(A): Choose from the two below:

A1: 30 dollars for sure

A2: 45 dollars with probability 0.8

(B): Choose from the two below:

B1: 30 dollars with probability 0.25
B2: 45 dollars with probability 0.2

Then the paradox is that one chooses A1 in (A) and B2 in (B).

Now, notice that the alternatives in (B) are equivalent to (C):

C1: Receive A1 with probability 0.25 and 0 dollars with probability 0.75
C2: Receive A2 with probability 0.25 and 0 dollars with probability 0.75

respectively.

Then, choosing B2 in (B) is choosing C2 in (C), but once lottery C2 resolves and you are about to receive A2, you go back to problem (A) and change your mind, saying "No, I want A1 now," which is a dynamic inconsistency.

Note, however, that the above argument presumes B1 and C1 are truly identical and B2 and C2 are truly identical. The lotteries in (C) are two-stage lotteries and those in (B) are one-stage lotteries, which are different physical objects per se.

For example, think of flipping a fair coin twice and giving 40 dollars if we see heads-heads, 30 if heads-tails, 20 if tails-heads and 10 if tails-tails. We can also think of throwing a four-face die and giving 40 dollars if we see 1, 30 if 2, 20 if 3 and 10 if 4. They are different physical objects. The expected utility theory assumes that the decision maker cares only about probability distributions over final outcomes.

To clarify what the expected utility theory is assuming implicitly, let us consider the following two-stage model, following Karni and Schmeidler [65].

At Period 0, the decision maker chooses over two-stage lotteries. The set of two-stage lotteries is given by $\Delta_S(\Delta_S(Z))$, the set of simple lotteries over $\Delta_S(Z)$, which is the set of simple lotteries over Z.

A two-stage simple lottery $l_0 \in \Delta_S(\Delta_S(Z))$ specifies a finite set $S(l_0) \subset \Delta_S(Z)$ and assigns each $l_1 \in S(l_0)$ a probability value $l_0(l_1) > 0$, so that $\sum_{l_1 \in S(l_0)} l_0(l_1) = 1$ holds.

Note that the set of one-stage lotteries $\Delta_S(Z)$ is seen as a subset of $\Delta_S(\Delta_S(Z))$, by seeing $l \in \Delta_S(Z)$ as a two-stage lottery $(\delta(z), l(z))_{z \in S(l)} \in \Delta_S(\Delta_S(Z))$ in which each realization $\delta(z)$ is a lottery degenerate on $z \in S(l)$.

Let $\succsim_0$ denote the Period-0 preference over $\Delta_S(\Delta_S(Z))$, the set of two-stage lotteries. Let $\succsim_1$ denote the Period-1 preference over $\Delta_S(Z)$, the set of one-stage lotteries.

Axiom 4.1 (Consequentialism) The Period-1 preference is defined independently of how a one-stage lottery is generated as a realization of some original two-stage lottery that was given in Period 0.

What it means will be better understood by seeing how it is violated in the later section, so please just leave it now.

Axiom 4.2 (Dynamic Consistency) For all two-stage lotteries $l_0, m_0 \in \Delta_S(\Delta_S(Z))$ given in the form $l_0 = (l_{1k}; \lambda_{0k})_{k=1}^n$ and $m_0 = (m_{1k}; \lambda_{0k})_{k=1}^n$, if $l_{1k} \succsim_1 m_{1k}$ for all $k = 1, \cdots, n$, then $l_0 \succsim_0 m_0$.

Axiom 4.3 (Reduction of Two-Stage Lotteries) For all $l_0 \in \Delta_S(\Delta_S(Z))$,

$$l_0 \sim_0 \sum_{l_1 \in S(l_0)} l_0(l_1) l_1.$$

Proposition 4.2 *If* $(\succsim_0, \succsim_1)$ *satisfy Consequentialism, Reduction of Two-stage Lotteries and Dynamic Consistency, then* $\succsim_0$ *satisfies Independence over* $\Delta_S(X)$.

Proof Pick any $l_0, m_0 \in \Delta_S(Z) \subset \Delta_S(\Delta_S(Z))$, and suppose $l_0 \succsim_0 m_0$. Pick any $q_0 \in \Delta_S(Z) \subset \Delta_S(\Delta_S(Z))$, then $q_0 \sim_0 q_0$. Then, because of Reduction of Two-Stage Lotteries we have $(l_0; 1) \succsim_0 (m_0; 1)$ and $(q_0; 1) \sim_0 (q_0; 1)$.

By Dynamic Consistency, $l_0 \succsim_1 m_0$ and $q_0 \sim_1 q_0$. Again by Dynamic Consistency, for all $\lambda \in [0, 1]$ it follows that $(l_0; \lambda, q_0; 1-\lambda) \succsim_0 (m_0; \lambda, q_0; 1-\lambda)$. By Reduction of Two-Stage Lotteries, we obtain $\lambda l_0 + (1-\lambda) q_0 \succsim_0 \lambda m_0 + (1-\lambda) q_0$. □

Thus, if want to allow a weakening of the independence axiom we have to give up either Consequentialism, Reduction of Two-Stage Lotteries (or so-called Timing Indifference) and Dynamic Consistency.

What do we mean by giving up consequentialism? Let me quote an example from Machina [76].

Example 4.6

Mom has a single indivisible item — a "treat" — which she can give to either daughter Abigail or son Benjamin. Assume that she is indifferent between Abigail getting the treat and Benjamin getting the treat, and strongly prefers either of these outcomes to the case where neither child gets it. However, in a violation of the precepts of expected utility theory, Mom strictly prefers a coin flip over either of these sure outcomes, and in particular, strictly prefers 1/2: 1/2 to any other pair of probabilities. This random allocation procedure would be straightforward, except that Benjie, who cut his teeth on Raiffa's classic Decision Analysis, behaves as follows:

> Before the coin is flipped, he requests a confirmation from Mom that, yes, she does strictly prefer a 50:50 lottery over giving the treat to Abigail. He gets her to put this in writing. Had he won the flip, he would have claimed the treat. As it turns out, he loses the flip. But as Mom is about to give the treat to Abigail, he reminds Mom of her preference for flipping a coin over giving it to Abigail (producing her signed statement), and demands that she flip again.

What would your Mom do if you tried to pull a stunt like this? She would undoubtedly say "You had your chance!" and refuse to flip the coin again. This is precisely what Mom does.

■

Machina continues, "By replying "You had your chance," Mom is reminding Benjamin of the existence of the snipped-off branch (the original 1/2 probability of B) and that her preferences are not separable, so the fact that nature could have gone down that branch still matters. Mom is rejecting the property of consequentialism—and, in my opinion, rightly so."

If we accept consequentialism, Benjie is right and Mom should flip a coin again, while it means Mom has to repeat flipping the coin forever.

On the other hand, saying "You had your chance" reveals that decision making after flipping the coin and seeing tails is dependent on having the plan that if it were heads Benjie would have won. In this sense we have to have a departure from consequentialism.

Chapter 5
Decision Under Uncertainty and Subjective Belief

When we are not given a probability distribution as an object as in the von-Neumann/Morgenstern expected utility theory, the decision maker has to form one subjectively in a certain way. While it will be something that is obtained by processing some objective information, it is ultimately a subjective one. Then, how can an outside observer identify such subjective belief from the decision maker's choices?

The subjective expected utility theory, due to Savage [91], deals with how a subjective belief is revealed from choices over bets.

5.1 Savage's Subjective Expected Utility Theory*

The basic idea is to compare between two bets,

> you win 100 dollars if Red Sox wins, nothing otherwise,

and

> you win 100 dollars if Celtics wins, nothing otherwise,

and to say that the decision maker believes Red Sox' winning is more likely than Celtics' winning when he chooses the first bet over the second.

By observing the decision maker's choices between bets, which induces a ranking over such bets, we can identify a ranking describing "likelihood" of events that is believed by him.

This is only a qualitative ranking stating that one event is believed to be more likely than another, and it is not clear if this allows a quantitative description by means of a probabilistic concept, stating how much an event is likely to happen one believes. Also it is not clear if such a probabilistic description is uniquely determined. The subjective expected utility theory due to Savage [91] identifies a set

T. Hayashi, *Decision Theory*, Monographs in Mathematical Economics 9,
https://doi.org/10.1007/978-981-95-2200-2_5

of conditions on preference over bets under which such a quantitative description is possible.

Let Ω denote the set of states of the world. For example, when we are talking about weather, the set would be $\Omega = \{\text{sunny}, \text{cloudy}, \text{rainy}, \text{snowy}, \cdots\}$, and when we are talking about temperature the set would be $\Omega = [0, \infty)$, where the temperature is in kelvins.

In subjective expected utility theory, the set of states Ω is **objectively fixed**. Although which one in Ω is more *probable* is taken to be a subjective matter, which states of the world are *possible* is taken to be objectively determined beforehand.

For example, the set of weather conditions $\Omega = \{\text{sunny}, \text{cloudy}, \text{rainy}, \text{snowy}, \cdots\}$ is supposed to describe all the objectively possible weather conditions. Now, however, let us consider the possibility that spears are falling from the heavens. In subjective expected utility theory, the fact that spears falling from the heavens is objectively impossible and the fact that the decision maker is unaware of such possibility as he cannot even imagine it are taken to be equivalent. Therefore, we are excluding the case that spears falling from the heavens should be considered at least as a possibility, while it may be likely or unlikely, and that the decision maker is nevertheless unaware of it.

Is it possible that an outside observer can identify if the decision maker is aware or unaware of some states of the world? It is easy to see that we cannot ask "are you aware of the possibility that spears falling from the heavens?," since the issue is then immediately over. Let me come to this issue in the chapter on information and knowledge.

Let X be the set of outcomes. An easy example of an outcome is an amount of money or goods, but we can think of more general and abstract outcomes, while it is important to note that outcomes are separated from states, as I shall come to below.

A mapping from the set of states Ω into the set of outcomes X is called an **act**. An act describes a functional statement saying "if this happens you get this, if that happens you get that, ..." It is our object of choice, and it gives formal description of a bet.

In particular, when an act $f : \Omega \to X$ has finite number of outcomes, that is, when $|f(\Omega)|$ is finite, we say that f is a **simple act**. Let $\mathcal{F}$ denote the set of simple acts. The preference relation $\succsim$ is defined over $\mathcal{F}$.

Let me emphasize that the functional relationship given by an act is an objective one. Thus, in applications, the decision maker is assumed to know how his physical or social action is objectively described as a mapping from each state into an outcome.

Also, we should pay attention to the assumption that states and outcomes are separated. What does it mean? Let us consider the following example from Savage [91], to understand.

Example 5.1 Consider that the set of "raw" physical actions is {Go to tennis, bring umbrella}. The uncertainty is about weather, which is either rain or shine. How do we model this?

It will be fine to take the set of states to be $\Omega = \{\text{rain, shine}\}$. The problem is the set of outcomes.

(a) Suppose we set $X = \{\text{tennis, umbrella}\}$. Then the evaluation of outcomes must be state-dependent. If it is sunny we would have tennis $\succ$ umbrella, and if it is rainy we would have tennis $\prec$ umbrella. As discussed later, when evaluation of outcomes is state-dependent we cannot have a well-defined concept of subjective belief which can be identifiable from choices.

(b) How about writing

$$X = \{\text{tennis under sunshine, tennis in the rain, umbrella under sunshine,} \\ \text{umbrella in the rain}\}?$$

Then, we have to think of an act for example that maps a state "rain" into an outcome "tennis under sunshine," which is absurd.

To avoid this kind of problem, the outcome should be taken to be easier to separate from states, such as monetary receipt and final consumption goods.

Having said the above remarks, we consider the following axioms.

P1 (Weak Order): $\succsim$ is complete and transitive.

P2 (Eventwise Separability (or Sure-Thing Principle)): For all $E \subset \Omega$ and $f, g, h, h' \in \mathcal{F}$,

$$\begin{bmatrix} f(\omega) \text{ if } \omega \in E \\ h(\omega) \text{ if } \omega \notin E \end{bmatrix} \succsim \begin{bmatrix} g(\omega) \text{ if } \omega \in E \\ h(\omega) \text{ if } \omega \notin E \end{bmatrix}$$

$$\Longrightarrow \begin{bmatrix} f(\omega) \text{ if } \omega \in E \\ h'(\omega) \text{ if } \omega \notin E \end{bmatrix} \succsim \begin{bmatrix} g(\omega) \text{ if } \omega \in E \\ h'(\omega) \text{ if } \omega \notin E \end{bmatrix}.$$

P2 has two roles. One is to ensure that subjective belief is additive. The other is to ensure that the independence axiom is met regarding lotteries that are induced by acts and the subjective belief obtained as a probability measure.

One might think that one axiom explaining two different things is "impure." Machina and Schmeidler [77] weaken P2 so that only the first role is played. We come to this later.

To introduce the third axiom we need one definition.

Definition 5.1 An event $E \subset \Omega$ is said to be *null* if any $f, g \in \mathcal{F}$ with $f = g$ on E^c yields $f \sim g$. Otherwise, E is said to be *non-null*.

P3 (Eventwise Monotonicity): For all $x, y \in X$ and any non-null event $E \subset \Omega$, and for all $g \in \mathcal{F}$,

$$\begin{bmatrix} x & \text{if } \omega \in E \\ g(\omega) & \text{if } \omega \notin E \end{bmatrix} \succsim \begin{bmatrix} y & \text{if } \omega \in E \\ g(\omega) & \text{if } \omega \notin E \end{bmatrix}$$

$$\Longleftrightarrow x \succsim y.$$

P3 guarantees state-independence of evaluation of outcomes. The subjective expected utility theory requires that evaluation of outcomes and subjective belief are separated. For example, suppose the decision maker chooses a bet "beer if Red Sox win and nothing otherwise" over another bet "beer if Celtics win and nothing otherwise." The theory requires an outside observer to infer from this observation that the decision maker believes Red Sox' winning is more likely than Celtics' winning. However, it does not make sense if the decision maker makes such a choice simply because he feels beer tastes better if Red Sox win.

P4 (Weak Comparative Probability): For all $A, B \subset \Omega$ and $x^*, x, y^*, y \in X$ with $x^* \succ x$ and $y^* \succ y$,

$$\begin{bmatrix} x^* & \text{if } \omega \in A \\ x & \text{if } \omega \notin A \end{bmatrix} \succsim \begin{bmatrix} x^* & \text{if } \omega \in B \\ x & \text{if } \omega \notin B \end{bmatrix}$$

$$\Longrightarrow \begin{bmatrix} y^* & \text{if } \omega \in A \\ y & \text{if } \omega \notin A \end{bmatrix} \succsim \begin{bmatrix} y^* & \text{if } \omega \in B \\ y & \text{if } \omega \notin B \end{bmatrix}.$$

Under **P4**, we can define **comparative likelihood** $\succsim^l$ in a well-defined manner. Define $A \succsim^l B$ by

$$\begin{bmatrix} x^* & \text{if } \omega \in A \\ x & \text{if } \omega \notin A \end{bmatrix} \succsim \begin{bmatrix} x^* & \text{if } \omega \in B \\ x & \text{if } \omega \notin B \end{bmatrix} \quad \text{for all } x^* \succ x.$$

Then **P4** guarantees that this definition does not depend on the choice of x^*, x.

P4 also requires, at another level, that evaluation of outcomes and subjective belief are separated. Again, suppose the decision maker chooses a bet "beer if Red Sox win and nothing otherwise" over another bet "beer if Celtics win and nothing otherwise." The theory requires an outside observer to infer from this observation that the decision maker believes Red Sox' winning is more likely than Celtics' winning. However, it does not make sense if the decision maker's choice is reversed when we replace beer with afternoon tea.

Note that **P1** guarantees that $\succsim^l$ is complete and transitive.

The next one will need no explanation, otherwise the theory is trivial.

P5 (Nondegeneracy): There exist $x, y \in X$ such that $x \succ y$.

P6 (Small Event Continuity): For all $f \succ g$ and $x \in X$, there exists a partition of Ω, $\{E_k\}_{k=1}^n$, such that

$$f \succ \begin{bmatrix} x & \text{if } \omega \in E_k \\ g(\omega) & \text{if } \omega \notin E_k \end{bmatrix}$$

and

$$\begin{bmatrix} x & \text{if } \omega \in E_k \\ f(\omega) & \text{if } \omega \notin E_k \end{bmatrix} \succ g$$

hold for all $k = 1, \cdots, n$.

P6 states that we can divide the set of states arbitrarily finely, which essentially says that the set is a continuum. For example, **P6** is not met when the set of states is finite. We come back to this point when we introduce the subjective expected utility theory due to Anscombe and Aumann.

Definition 5.2 A set function $p : 2^\Omega \to \mathbb{R}$ is said to be a finitely additive probability measure if

$$p(\emptyset) = 0, \quad p(\Omega) = 1$$

and

$$A \subset B \implies p(A) \leq p(B)$$

hold for all $A, B \subset \Omega$, and for all $A_1, \cdots, A_n \subset \Omega$ with $A_j \cap A_k = \emptyset$ for every $j, k = 1, \cdots, n$,

$$p\left(\bigcup_{k=1}^n A_k\right) = \sum_{k=1}^n p(A_k).$$

Say that p, a probability measure over Ω is **convex-ranged** if for all E with $p(E) > 0$ and for all $0 < \alpha < 1$ there is an event $\exists E^* \subset E$ such that $p(E^*) = \alpha p(E)$.

Now the subjective expected utility representation theorem due to Savage is written as follows.

Theorem 5.1 *The following two statements are equivalent:*

(a) $\succsim$ *satisfies P1-6.*
(b) There exist a convex-ranged finitely additive probability measure p *over* Ω *and a non-constant function* $u : X \to \mathbb{R}$ *such that* $\succsim$ *is represented by* $U : \mathcal{F} \to \mathbb{R}$

in the form

$$U(f) = \sum_{x \in f(\Omega)} u(x)\ p(f^{-1}(x)).$$

Moreover, if another such pair (q, v) *forms a representation in the same form then* $p = q$ *and there exist constants* $a > 0$ *and* b *such that* $v = au + b$.

Remark 5.1 To allow non-simple acts, Savage [91] and Fishburn [35] imposed an additional axiom called P7, which guarantees that the function u is bounded. It has been shown, however, by Hartmann [56], that P3 is redundant once P7 is added.

Proof of the Savage Theorem

Honestly, I cannot do anything better than reproduce the proof by Fishburn [35]. Nevertheless, I think it is worth doing in the current context for the sake of being self-contained. Also I tried to fill several steps which are taken to be straightforward in Fishburn, and tried to make it more accessible to introductory learners.

Step 1: Show that the comparative likelihood relation $\succsim^l$ satisfies the condition of **qualitative probability** below.

F1: For all $A \subset \Omega$ it holds $A \succsim^l \emptyset$.
F2: $\Omega \succ^l \emptyset$
F3: The relation $\succsim^l$ is complete and transitive.
F4: If $A \cap C = B \cap C = \emptyset$, then it holds $A \succsim^l B \Longleftrightarrow A \cup C \succsim^l B \cup C$.

Proof We prove F4 and the rest are left as exercises. Pick any A, B, C with $A \cap C = B \cap C = \emptyset$, pick any $x^*, x \in X$ with $x^* \succ x$, and suppose

$$\begin{bmatrix} x^* \text{ if } \omega \in A \\ x \ \text{ if } \omega \notin A \end{bmatrix} \succsim \begin{bmatrix} x^* \text{ if } \omega \in B \\ x \ \text{ if } \omega \notin B \end{bmatrix}.$$

This is written as

$$\begin{bmatrix} x^* \text{ if } \omega \in A \\ x \ \ \text{ if } \omega \in C \\ x \ \ \text{ if } \omega \notin A \cup C \end{bmatrix} \succsim \begin{bmatrix} x^* \text{ if } \omega \in B \\ x \ \ \text{ if } \omega \in C \\ x \ \ \text{ if } \omega \notin B \cup C \end{bmatrix}.$$

By P2, the above is equivalent to

$$\begin{bmatrix} x^* \text{ if } \omega \in A \\ x^* \text{ if } \omega \in C \\ x \ \ \text{ if } \omega \notin A \cup C \end{bmatrix} \succsim \begin{bmatrix} x^* \text{ if } \omega \in B \\ x^* \text{ if } \omega \in C \\ x \ \ \text{ if } \omega \notin B \cup C \end{bmatrix},$$

which is nothing but

$$\begin{bmatrix} x^* \text{ if } \omega \in A \cup C \\ x \ \text{ if } \omega \notin A \cup C \end{bmatrix} \succsim \begin{bmatrix} x^* \text{ if } \omega \in B \cup C \\ x \ \text{ if } \omega \notin B \cup C \end{bmatrix}.$$

□

The above properties are necessary for the relation $\succsim^l$ to be represented by a unique probability measure, but **it is not sufficient**. It becomes sufficient together with the following property.

F5: If $A \succ^l B$, there is a partition $\{E_k\}_{k=1}^n$ of Ω such that $A \succ^l B \cup E_k$ holds for all $k = 1, \cdots, n$.

Proof Suppose

$$\begin{bmatrix} x^* \text{ if } \omega \in A \\ x \ \text{ if } \omega \notin A \end{bmatrix} \succ \begin{bmatrix} x^* \text{ if } \omega \in B \\ x \ \text{ if } \omega \notin B \end{bmatrix}.$$

Then, by P6 there is a partition $\{E_k\}_{k=1}^n$ of Ω such that

$$\begin{bmatrix} x^* \text{ if } \omega \in A \\ x \ \text{ if } \omega \notin A \end{bmatrix} \succ \begin{bmatrix} x^* \text{ if } \omega \in E_k \\ x^* \text{ if } \omega \in B \setminus E_k \\ x \ \text{ if } \omega \notin B \cup E_k \end{bmatrix}$$

holds for all $k = 1, \cdots, n$. This is equivalent to saying that

$$\begin{bmatrix} x^* \text{ if } \omega \in A \\ x \ \text{ if } \omega \notin A \end{bmatrix} \succ \begin{bmatrix} x^* \text{ if } \omega \in B \cup E_k \\ x \ \text{ if } \omega \notin B \cup E_k \end{bmatrix}$$

holds for all $k = 1, \cdots, n$. □

Step 2: Show that **F1–5** are equivalent to the existence of (finitely additive) probability measure p over 2^Ω such that

$$A \succsim^l B \iff p(A) \geq p(B)$$

holds for all $A, B \subset \Omega$.

For this purpose, we prove the following nine properties:

C1: If $A \supset B$, then $\emptyset \precsim^l B \precsim^l A \precsim^l \Omega$.

C2: If $A \sim^l B$ and $B \cap C = \emptyset$, then it holds $A \cup C \precsim^l B \cup C$.
Also, if $A \prec^l B$ and $B \cap C = \emptyset$, then $A \cup C \prec^l B \cup C$.

C3: If $A \sim^l B, C \sim^l D$ and $B \cap D = \emptyset$, then $A \cup C \precsim^l B \cup D$.
Also, if $A \precsim^l B, C \prec^l D$ and $B \cap D = \emptyset$, then $A \cup C \prec^l B \cup D$.

C4: If $A \sim^l B, C \sim^l D$ and $A \cap C = B \cap D = \emptyset$, then $A \cup C \sim^l B \cup D$.

C5: If $A \succ^l \emptyset$, then there is a partition B, C of A such that $B, C \succ^l \emptyset$.

C6: If A, B, C are mutually disjoint and $A \precsim^l B \prec^l A \cup C$, then there is $D \subset C$ with $D \succ^l \emptyset$ such that $B \cup D \prec^l A \cup (C \setminus D)$ holds.

C7: If $A, B \succ^l \emptyset$ and $A \cap B = \emptyset$, then there is a partition C, D of B such that $C \precsim^l D \precsim^l A \cup C$ holds.

C8: If $A \succ^l \emptyset$, then there is a partition B, C of A such that $B \sim^l C$ holds.

C9: If $A \succ^l \emptyset$, then for all n there is a partition $\{A_i^n\}_{i=1}^{2^n}$ of A such that $A_i^n \sim^l A_j^n$ holds for all $1 \leq i, j \leq 2^n$.

Proof **C1**: Suppose $A \supset B$. Then, for arbitrary $x^*, x \in X$ with $x^* \succ x$, from P3

$$\begin{bmatrix} x^* \text{ if } \omega \in B \\ x^* \text{ if } \omega \in A \setminus B \\ x \ \ \text{ if } \omega \notin A \end{bmatrix} \succsim \begin{bmatrix} x^* \text{ if } \omega \in B \\ x \ \ \text{ if } \omega \in A \setminus B \\ x \ \ \text{ if } \omega \notin A \end{bmatrix},$$

which is equivalent to

$$\begin{bmatrix} x^* \text{ if } \omega \in A \\ x \ \ \text{ if } \omega \notin A \end{bmatrix} \succsim \begin{bmatrix} x^* \text{ if } \omega \in B \\ x \ \ \text{ if } \omega \notin B \end{bmatrix}.$$

C2: Suppose $A \sim^l B$ and $B \cap C = \emptyset$. Then by definition we have

$$\begin{bmatrix} x^* \text{ if } \omega \in A \\ x \ \ \text{ if } \omega \in A^c \end{bmatrix} \sim \begin{bmatrix} x^* \text{ if } \omega \in B \\ x \ \ \text{ if } \omega \in B^c \end{bmatrix}.$$

Since $C \setminus A \subset A^c \cap B^c$, by P2 we have

$$\begin{bmatrix} x^* \text{ if } \omega \in A \\ x \ \ \text{ if } \omega \in A^c \setminus (C \setminus A) \\ x^* \text{ if } \omega \in C \setminus A \end{bmatrix} \sim \begin{bmatrix} x^* \text{ if } \omega \in B \\ x \ \ \text{ if } \omega \in B^c \setminus (C \setminus A) \\ x^* \text{ if } \omega \in C \setminus A \end{bmatrix},$$

meaning $A \cup (C \setminus A) \sim^l B \cup (C \setminus A)$. Since $C \setminus A \subset C$, we have $B \cup (C \setminus A) \subset B \cup C$. Hence by C1 we obtain $B \cup (C \setminus A) \precsim^l B \cup C$, implying $A \cup (C \setminus A) \precsim^l B \cup C$. Since $A \cup (C \setminus A) = A \cup C$, we obtain the conclusion. Similarly for the second claim.

C3: Suppose $A \sim^l B$, $C \sim^l D$ and $B \cap D = \emptyset$. Since $B \cap (C \setminus B) = \emptyset$, from C2 and $A \sim^l B$ we have $A \cup (C \setminus B) \precsim^l B \cup (C \setminus B) = B \cup C$. Also, since $B \cap D = \emptyset$ implies $(B \setminus C) \cap D = \emptyset$, from C2 and $C \sim^l D$ we have $B \cup C = C \cup (B \setminus C) \precsim^l D \cup (B \setminus C)$. Thus by transitivity we obtain $A \cup (C \setminus B) \precsim^l D \cup (B \setminus C)$. Because $(B \cap C) \cap (D \cup (B \setminus C)) = \emptyset$, from C2 we have $(A \cup (C \setminus B)) \cup (B \cap C) \precsim^l (D \cup (B \setminus C)) \cup (B \cap C)$. Now the conclusion follows from $(A \cup (C \setminus B)) \cup (B \cap C) = A \cup C$ and $(D \cup (B \setminus C)) \cup (B \cap C) = D \cup B$. Similarly for the second claim.

C4: Suppose $A \sim^l B$, $C \sim^l D$ and $A \cap C = B \cap D = \emptyset$. Then by C3 we have $A \cup C \precsim^l B \cup D$ and $B \cup D \precsim^l A \cup C$, which imply $A \cup C \sim^l B \cup D$.

C5: Suppose $A \succ^l \emptyset$. Then by F5 there is a partition of Ω, denoted by $\{D_1, \cdots, D_n\}$, such that $D_k \prec^l A$ holds for all $k = 1, \cdots, n$. By C1, $D_k \cap A \precsim^l D_k$ holds for all $k = 1, \cdots, n$. By transitivity, $D_k \cap A \prec^l A$ holds for all $k = 1, \cdots, n$.

Suppose $D_k \cap A \sim^l \emptyset$ is true for all for all $k = 1, \cdots, n$. Then by repeatedly applying C4 we have $\bigcup_{k=1}^n (D_k \cap A) = A \sim^l \emptyset$, a contradiction. Now suppose there is only one k, let's say $k = 1$, such that $D_1 \cap A \succ^l \emptyset$. Then for all $k = 2, \cdots, n$ we must have $D_k \cap A \sim^l \emptyset$. Hence again by applying C4 repeatedly we obtain $A \sim^l D_1 \cap A$, which contradicts to $D_1 \cap A \prec^l A$. Thus there are at least two k such that $D_k \cap A \succ^l \emptyset$.

C6: Suppose A, B, C are mutually disjoint and $A \precsim^l B \prec^l A \cup C$. If $C \sim^l \emptyset$, then from F4 we have $A \cup C \sim^l A$, which contradicts the assumption. Hence we must have $C \succ^l \emptyset$. Then, from F5 there is a partition of Ω, denoted by $\{E_k\}_{k=1}^n$, such that $B \cup E_k \prec^l A \cup C$ holds for all $k = 1, \cdots, n$.

Now if $E_k \cap C \sim^l \emptyset$ for all $k = 1, \cdots, n$, then repeatedly applying C4 we obtain $\bigcup_{k=1}^n (E_k \cap C) = C \sim^l \emptyset$, which contradicts the preceding claim. Hence there is at least one k such that $E_k \cap C \succ^l \emptyset$. Let's say this k is 1, and let $D_1 = E_1 \cap C$. Then we have $D_1 \succ^l \emptyset$, $D_1 \subset C$ and $B \cup D_1 \prec^l A \cup C$.

From C5 and F3, D_1 can be partitioned to D and D' such that $\emptyset \prec^l D \precsim^l D'$. Then we have $B \cup D \cup D' = B \cup D_1 \prec^l A \cup C = A \cup (C \setminus D) \cup D$. From F4 this implies $B \cup D' \prec^l A \cup (C \setminus D)$. On the other hand, because of $D \precsim^l D'$ and F4 we have $B \cup D \precsim^l B \cup D'$. Thus we obtain $B \cup D \prec^l A \cup (C \setminus D)$.

C7: Suppose $A, B \succ^l \emptyset$ and $A \cap B = \emptyset$. If $B \precsim^l A$ then the conclusion immediately follows from C5. Hence assume $A \prec^l B$. Then, from F5 there is a partition of Ω, denoted by $\{E_k\}_{k=1}^n$, such that $E_k \prec^l A$ holds for all $k = 1, \cdots, n$. Write $G_k = B \cap E_k$ for each $k = 1, \cdots, n$. Then $G_k \prec^l A$ holds for all $k = 1, \cdots, n$. Without loss of generality, assume $G_1 \precsim^l G_2 \precsim^l \cdots \precsim^l G_n$.

Now consider m such that $\bigcup_{k=1}^m G_k \precsim^l \bigcup_{k=m+1}^n G_k \precsim^l \bigcup_{k=1}^{m+1} G_k$, and write $C = \bigcup_{k=1}^m G_k$, $D = \bigcup_{k=m+1}^n G_k$. Then we have $C \precsim^l D \precsim^l C \cup G_{m+1}$. Because $G_{m+1} \prec^l A$, $G_{m+1} \cap C = \emptyset$ and $A \cap C = \emptyset$, from F4 and F3 we obtain $D \prec^l C \cup A$.

C8: Suppose $A \succ^l \emptyset$. By C5, there exists a partition of A, denoted by B, C, such that $B, C \succ^l \emptyset$. Assume without loss of generality that $B \precsim^l C$. By C7, there is a partition of C, denoted by C_1, D_1, such that $D_1 \precsim^l C_1 \precsim^l B \cup D_1$, hence $B \precsim^l C_1 \cup D_1$. Now let $B_1 = B$, then B_1, C_1, D_1 form a partition of A such that $B_1 \precsim^l C_1 \cup D_1$ and $C_1 \precsim^l B_1 \cup D_1$ hold.

The conclusion is that C8 would have been obtained if at least one of these two relations holds with $\sim^l$, hence we assume $B_1 \prec^l C_1 \cup D_1$ and $C_1 \prec^l B_1 \cup D_1$. Without loss, assume $B_1 \precsim^l C_1$.

Then, because $D_1 \succ^l \emptyset$, from $C_1 \prec^l B_1 \cup D_1$ and C6 there exists some $C^2 \subset D_1$ such that $C^2 \succ^l \emptyset$ and $C_1 \cup C^2 \prec^l B_1 \cup (D_1 \setminus C^2)$ hold.

Then $D_1 \setminus C^2 \succ^l \emptyset$ holds, for, if $D_1 \setminus C^2 \sim^l \emptyset$ then combined with $C^2 \succ^l \emptyset$ it implies $B_1 \cup (D_1 \setminus C^2) \sim^l B_1 \precsim^l C_1 \prec^l C_1 \cup C^2$, which contradicts the preceding argument.

Hence by C7, $D_1 \setminus C^2$ is partitioned into B^2 and D_2 such that $B^2 \precsim^l D_2 \precsim^l C^2 \cup B^2$ holds. Since $B_1 \precsim^l C_1$ holds by assumption, by repeatedly applying C3 we obtain $B_1 \cup B^2 \precsim^l C_1 \cup D_2 \prec^l C_1 \cup D_2 \cup C^2$.

Now let

$$B_2 = B_1 \cup B^2, \quad C_2 = C_1 \cup C^2.$$

Then B_2, C_2, D_2 make a partition of A satisfying:

1. $B_2 \prec^l C_2 \cup D_2$ and $C_2 \prec^l B_2 \cup D_2$,
2. $B_1 \subset B_2, C_1 \subset C_2, D_2 \subset D_1$,
3. $\emptyset \prec^l D_2 \precsim^l D_1 \setminus D_2$,

where the third condition follows from $D_1 \setminus D_2 = D_1 \setminus ((D_1 \setminus C^2) \setminus B^2) = C^2 \cup B^2$.

By repeating this, we obtain a sequence of partitions of A, denoted by $\{B_n, C_n, D_n\}$, which satisfies the following for all n:

1. $B_n \prec^l C_n \cup D_n$ and $C_n \prec^l B_n \cup D_n$,
2. $B_n \subset B_{n+1}, C_n \subset C_{n+1}, D_{n+1} \subset D_n$,
3. $\emptyset \prec^l D_{n+1} \precsim^l D_n \setminus D_{n+1}$.

Note that from Property 3 above, $E_1, E_2 \prec^l D_{n+1}$ implies $E_1 \cup E_2 \prec^l D_n$ for arbitrary E_1, E_2, because $E_1, E_2 \precsim^l D_n \setminus D_{n+1}$ holds.

Then, for arbitrary G with $G \succ^l \emptyset$, by taking sufficiently large n we have $D_n \prec^l G$. To show this, suppose $G \precsim^l D_n$. Then by F5 we can take a partition of Ω, denoted by $\{E_1, \cdots, E_m\}$, such that $E_k \prec^l D_n$ holds for all $k = 1, \cdots, m$. From the preceding argument $E_1 \cup E_2, E_3 \cup E_4, \cdots \prec^l D_{n-1}$, and likewise $\cup_{k=1}^4 E_k, \cup_{k=5}^8 E_k, \cdots \prec^l D_{n-2}$, and by repeating this we obtain $\cup_{k=1}^m E_k = \Omega \prec^l D_1$, which is a contradiction. This leads to $\bigcap_{n=1}^\infty D_n \sim^l \emptyset$. For, if $\bigcap_{n=1}^\infty D_n \succ^l \emptyset$, then by taking sufficiently large m we have $D_m \prec^l \bigcap_{n=1}^\infty D_n$, which contradicts $\bigcap_{n=1}^\infty D_n \subset D_m$.

Now we show that

$$B = \bigcup_{n=1}^\infty B_n, \quad C = \left(\bigcup_{n=1}^\infty C_n\right) \cup \left(\bigcap_{n=1}^\infty D_n\right)$$

make a partition of A such that $B \sim^l C$.

It is clear from the definition that $B, C \subset A$ and $B \cap C = \emptyset$. Also, since $\bigcap_{n=1}^\infty D_n \sim^l \emptyset$, $C \sim^l \bigcup_{n=1}^\infty C_n$.

Now suppose $B \prec^l C$. Then since $C \sim^l \bigcup_{n=1}^\infty C_n$ we have $B \prec^l \bigcup_{n=1}^\infty C_n$. By C6, there is $G \subset \bigcup_{n=1}^\infty C_n$ with $G \succ^l \emptyset$ such that

$$B \cup G \prec^l \left(\bigcup_{n=1}^\infty C_n\right) \setminus G$$

holds.

Since $B \cap G = \emptyset$ and $B_n \precsim^l B$ hold for all n (from $B_n \subset B$), by F4 it follows that

$$B_n \cup G \precsim^l B \cup G$$

Since $D_m \prec^l G$ holds for sufficiently large m, again by F4 it follows that

$$B_m \cup D_m \prec^l B_m \cup G.$$

On the other hand, for sufficiently large m, $D_m \cap \left(\bigcup_{n=1}^{\infty} C_n\right) \prec^l G$, hence by C3 we have

$$\left(\bigcup_{n=1}^{\infty} C_n\right) \setminus G \precsim^l \left(\bigcup_{n=1}^{\infty} C_n\right) \setminus D_m.$$

Since $\left(\bigcup_{n=1}^{\infty} C_n\right) \setminus D_m \subset C_m$, by C1 we obtain $\left(\bigcup_{n=1}^{\infty} C_n\right) \setminus D_m \precsim^l C_m$. Summing up, by transitivity we obtain

$$B_m \cup D_m \prec^l C_m$$

for sufficiently large m, this contradicts Property 1 of the sequence $\{B_n, C_n, D_n\}$. Thus we have shown $B \succsim^l C$.

We can show $B \precsim^l C$ similarly, and obtain $B \sim^l C$.

C9: Shown by repeating C8. □

From the above argument, Ω is partitioned into 2^n events for arbitrary n which are subjectively equally likely and denoted by $\mathcal{E}^n = \{E_i^n\}_{i=1}^{2^n}$, in the way that for $k = 1, \cdots, 2^n$ the event E_k^n is again partitioned into $E_{2k-1}^{n+1}, E_{2k}^{n+1} \in \mathcal{E}^{n+1}$, which are subjectively equally likely.

Define the set function $p : 2^{\Omega} \to [0, 1]$ as follows. When $A \sim^l \Omega$, set $p(A) = 1$. When $A \prec^l \Omega$, first define

$$\underline{k}(A, n) = \max\left\{k : k \leq 2^n,\ E_1^n, \cdots, E_k^n \in \mathcal{E}^n,\ \bigcup_{i=1}^{k} E_i^n \precsim^l A\right\},$$

which is the maximal number of subjectively equally likely events taken from $\mathcal{E}^n$ that approximate A from below. Then, each of these $\underline{k}(A, n)$ events $E_1^n, \cdots, E_{\underline{k}(A,n)}^n \in \mathcal{E}^n$ is again partitioned into two subjectively equally likely events and they belong to $\mathcal{E}^{n+1} = \{E_i^n\}_{i=1}^{2^{n+1}}$. Then these $2\underline{k}(A, n)$ events, denoted by $E_1^{n+1}, \cdots, E_{2\underline{k}(A,n)}^{n+1}$, satisfy $\bigcup_{i=1}^{2\underline{k}(A,n)} E_i^{n+1} \sim^l \bigcup_{i=1}^{\underline{k}(A,n)} E_i^n \precsim^l A$. Hence by the definition of maximum we have $\underline{k}(A, n+1) \geqq 2\underline{k}(A, n)$. Thus the sequence $\left\{\frac{\underline{k}(A,n)}{2^n}\right\}$ is non-decreasing and bounded above, implying that it converges.

Likewise, define

$$\overline{k}(A,n) = \min\left\{k : k \leq 2^n,\ E_1^n, \cdots, E_k^n \in \mathcal{E}^n,\ \bigcup_{i=1}^{k} E_i^n \succsim^l A\right\},$$

which is the minimal number of subjectively equally likely events taken from $\mathcal{E}^n$ that approximate A from above. Then, each of these $\overline{k}(A,n)$ events $E_1^n, \cdots, E_{\overline{k}(A,n)}^n \in \mathcal{E}^n$ is again partitioned into two subjectively equally likely events and they belong to $\mathcal{E}^{n+1} = \{E_i^n\}_{i=1}^{2^{n+1}}$. Then these $2\overline{k}(A,n)$ events, denoted by $E_1^{n+1}, \cdots, E_{2\overline{k}(A,n)}^{n+1}$, satisfy $\bigcup_{i=1}^{2\overline{k}(A,n)} E_i^{n+1} \sim^l \bigcup_{i=1}^{\overline{k}(A,n)} E_i^n \succsim^l A$. Hence by the definition of minimum we have $\overline{k}(A,n+1) \leqq 2\overline{k}(A,n)$. Thus the sequence $\left\{\frac{\overline{k}(A,n)}{2^n}\right\}$ is non-increasing and bounded below, implying that it converges.

Note that $0 \leqq \overline{k}(A,n) - \underline{k}(A,n) \leqq 1$ holds for all n, hence we have

$$\lim_{n\to\infty} \frac{\underline{k}(A,n)}{2^n} = \lim_{n\to\infty} \frac{\overline{k}(A,n)}{2^n}.$$

Thus we define

$$p(A) = \lim_{n\to\infty} \frac{\underline{k}(A,n)}{2^n} = \lim_{n\to\infty} \frac{\overline{k}(A,n)}{2^n}.$$

We now show that the qualitative probability is represented by a finitely additive probability measure with convex range.

Lemma 5.1 *The set function p obtained above is a finitely additive probability measure over* (Ω, Σ) *and has a convex range, and satisfies*

$$A \precsim^l B \iff p(A) \leqq p(B)$$

for all $A, B \subset \Omega$.

Proof It is clear that $p(\emptyset) = 0$, $p(\Omega) = 1$ and $p(A) \geq 0$ for all $A \subset \Omega$.

To show that it is finitely additive, suppose $A \cap B = \emptyset$. Then, without loss of generality, assume we can take $E_1, \cdots, E_{\underline{k}(A,n)}, E_{\underline{k}(A,n)+1}, E_{\underline{k}(A,n)+\underline{k}(B,n)}$ such that $\bigcup_{i=1}^{\underline{k}(A,n)} E_i^n \precsim^l A$ and $\bigcup_{i=\underline{k}(A,n)+1}^{\underline{k}(A,n)+\underline{k}(B,n)} E_i^n \precsim^l B$. Since $\bigcup_{i=1}^{\underline{k}(A,n)+\underline{k}(B,n)} E_i^n \precsim^l A \cup B$ we have

$$\underline{k}(A \cup B, n) \geq \underline{k}(A,n) + \underline{k}(B,n).$$

Hence

$$\frac{\underline{k}(A \cup B, n)}{2^n} \geq \frac{\underline{k}(A,n)}{2^n} + \frac{\underline{k}(B,n)}{2^n}$$

holds for all n. In the limit, we obtain

$$p(A \cup B) \geq p(A) + p(B)$$

in the limit.

Likewise, without loss assume we can take $E_1, \cdots, E_{\overline{k}(A,n)}, E_{\overline{k}(A,n)+1}, E_{\overline{k}(A,n)+\overline{k}(B,n)}$ such that $\bigcup_{i=1}^{\overline{k}(A,n)} E_i^n \succsim^l A$ and $\bigcup_{i=\overline{k}(A,n)+1}^{\overline{k}(A,n)+\overline{k}(B,n)} E_i^n \succsim^l B$.

Since $\bigcup_{i=1}^{\overline{k}(A,n)+\overline{k}(B,n)} E_i^n \succsim^l A \cup B$ we have

$$\overline{k}(A \cup B, n) \leq \overline{k}(A, n) + \overline{k}(B, n).$$

Hence

$$\frac{\overline{k}(A \cup B, n)}{2^n} \leq \frac{\overline{k}(A, n)}{2^n} + \frac{\overline{k}(B, n)}{2^n}$$

holds for all n. In the limit, we obtain

$$p(A \cup B) \leq p(A) + p(B)$$

in the limit.

Next we show

$$A \succsim^l B \iff p(A) \geq p(B).$$

Suppose $A \succsim^l B$ first, then because $\bigcup_{i=1}^k E_i^n \precsim^l B$ implies $\bigcup_{i=1}^k E_i^n \precsim^l A$, $\underline{k}(A, n) \geq \underline{k}(B, n)$ holds for all n. Hence we obtain $p(A) \geq p(B)$ in the limit.

Now suppose $A \succ^l B$. Then for sufficiently large n there exist $E_1^n, \cdots, E_k^n, E_{k+1}^n \in \mathcal{E}^n$ such that

$$A \succsim^l \bigcup_{i=1}^{k+1} E_i^n > \bigcup_{i=1}^{k} E_i^n \succsim^l B$$

and we obtain $\underline{k}(A, n) > \overline{k}(B, n)$ for all such n. Since the sequence $\left\{\frac{\underline{k}(A,n)}{2^n}\right\}$ is non-decreasing and the sequence $\left\{\frac{\overline{k}(B,n)}{2^n}\right\}$ is non-increasing, we obtain $p(A) > p(B)$ in the limit.

Finally, we show that p has a convex range. Since it is trivial if $A \sim^l \emptyset$, let us assume $A \succ^l \emptyset$. By C9, for all n there is a partition of A, denoted by $\{A_i^n\}_{i=1}^{2^n}$, such that $A_i^n \sim^l A_j^n$ for all $1 \leq i, j \leq 2^n$.

Let

$$l(\lambda, n) = \max\left\{l : \frac{l}{2^n} < \lambda\right\},$$

then from the definition of maximum,

$$p\left(\bigcup_{i=1}^{l(\lambda,n)} A_i^n\right) + \frac{1}{2^n} p(A) = \frac{l(\lambda, n) + 1}{2^n} p(A) \geq \lambda p(A).$$

Also, let

$$m(\lambda, n) = \min\left\{m : \frac{2^n - m}{2^n} < 1 - \lambda\right\},$$

then from the definition of minimum,

$$p\left(\bigcup_{i=m(\lambda,n)+1}^{2^n} A_i^n\right) + \frac{1}{2^n} p(A) = \frac{2^n - (m(\lambda, n) - 1)}{2^n} p(A) \geq (1 - \lambda) p(A).$$

Now consider increasing sequences of sets $\{\bigcup_{i=1}^{l(\lambda,n)} A_i^n\}$ and $\{\bigcup_{i=m(\lambda,n)+1}^{2^n} A_i^n\}$, then we obtain

$$p\left(\bigcup_{n=1}^{\infty} \bigcup_{i=1}^{l(\lambda,n)} A_i^n\right) \geq \lambda p(A)$$

and

$$p\left(\bigcup_{n=1}^{\infty} \bigcup_{i=m(\lambda,n)+1}^{2^n} A_i^n\right) \geq (1 - \lambda) p(A).$$

Because $\bigcup_{i=1}^{l(\lambda,n)} A_i^n, \bigcup_{i=m(\lambda,n)+1}^{2^n} A_i^n \subset A$ and $\bigcup_{i=1}^{l(\lambda,n)} A_i^n \cap \bigcup_{i=m(\lambda,n)+1}^{2^n} A_i^n = \emptyset$ for all n, by additivity of p,

$$\begin{aligned} p\left(\bigcup_{i=1}^{l(\lambda,n)} A_i^n\right) + p\left(\bigcup_{i=m(\lambda,n)+1}^{2^n} A_i^n\right) &= p\left(\bigcup_{i=1}^{l(\lambda,n)} A_i^n \cup \bigcup_{i=m(\lambda,n)+1}^{2^n} A_i^n\right) \\ &\leq p(A) \end{aligned}$$

holds for all n. Hence in the limit

$$p\left(\bigcup_{n=1}^{\infty}\bigcup_{i=1}^{l(\lambda,n)} A_i^n\right) + p\left(\bigcup_{n=1}^{\infty}\bigcup_{i=m(\lambda,n)+1}^{2^n} A_i^n\right) \leq p(A).$$

Combined with the previous claim we obtain

$$p\left(\bigcup_{n=1}^{\infty}\bigcup_{i=1}^{l(\lambda,n)} A_i^n\right) = \lambda p(A)$$

and

$$p\left(\bigcup_{n=1}^{\infty}\bigcup_{i=m(\lambda,n)+1}^{2^n} A_i^n\right) = (1-\lambda)p(A).$$

□

Step 3: Given the above-obtained probability measure p over Ω and a simple act f, define a simple lottery $\Psi_{p,f}$ over X by

$$\Psi_{p,f}(Y) = p(f^{-1}(Y)),\ Y \subset X.$$

Then we show that $\Psi_{p,f} = \Psi_{p,g}$ implies $f \sim g$ for all simple acts f, g.

Proof Pick arbitrary simple acts f, g and suppose $\Psi_{p,f} = \Psi_{p,g}$. Below we prove by induction.

The claim is trivial when $|f(\Omega)| = |g(\Omega)| = 1$.

Suppose $|f(\Omega)| = |g(\Omega)| = 2$, and let

$$f = \begin{array}{c|cc} & E_1 & E_2 \\ \hline D_1 & x_1 & x_1 \\ D_2 & x_2 & x_2 \end{array}, \quad g = \begin{array}{c|cc} & E_1 & E_2 \\ \hline D_1 & x_1 & x_2 \\ D_2 & x_1 & x_2 \end{array},$$

where $D_1 \sim^l E_1$ and $D_2 \sim^l E_2$.

Then, since $D_1 \setminus E_1 \sim^l E_1 \setminus D_1$ it follows that

$$\begin{array}{c|cc} & E_1 & E_2 \\ \hline D_1 & x_2 & x_1 \\ D_2 & x_2 & x_2 \end{array} \sim \begin{array}{c|cc} & E_1 & E_2 \\ \hline D_1 & x_2 & x_2 \\ D_2 & x_1 & x_2 \end{array}.$$

Now replace the value at $D_1 \cap E_1$ in both sides from x_2 to x_1 then they are f and g respectively, which implies $f \sim g$.

Now suppose the claim is true until $|f(\Omega)| = |g(\Omega)| = n - 1$, and show it for the case of $|f(\Omega)| = |g(\Omega)| = n$.

Let

$$f = \begin{bmatrix} x_1 & D_1 \\ x_2 & D_2 \\ \vdots & \vdots \\ x_{n-1} & D_{n-1} \\ x_n & D_n \end{bmatrix}, \quad g = \begin{bmatrix} x_1 & E_1 \\ x_2 & E_2 \\ \vdots & \vdots \\ x_{n-1} & E_{n-1} \\ x_n & E_n \end{bmatrix},$$

where $D_k \sim^l E_k$ holds for all $k = 1, \cdots, n$.

Since $D_n \setminus E_n \sim E_n \setminus D_n$, from convex-rangedness of p there is a partition of $D_n \setminus E_n$, denoted by $F_1, \cdots, F_{n-1}$, such that

$$F_k \sim^l D_k \cap E_n$$

holds for all $k = 1, \cdots, n-1$.

Now rewrite f as

$$f = \frac{\begin{array}{ccccc} F_1 & F_2 & \cdots & F_{n-1} & \\ \hline x_n & x_n & \cdots & x_n & \end{array}}{\begin{array}{cccccc} D_1 \cap E_n & D_2 \cap E_n & \cdots & D_{n-1} \cap E_n & D_n \cap E_n & D_n^c \cap E_n^c \\ \hline x_1 & x_2 & \cdots & x_{n-1} & x_n & f \end{array}}$$

and exchange its value at F_1, that is x_n, and the value at $D_1 \cap E_n$, that is x_1, and let

$$f_1 = \frac{\begin{array}{ccccc} F_1 & F_2 & \cdots & F_{n-1} & \\ \hline x_1 & x_n & \cdots & x_n & \end{array}}{\begin{array}{cccccc} D_1 \cap E_n & D_2 \cap E_n & \cdots & D_{n-1} \cap E_n & D_n \cap E_n & D_n^c \cap E_n^c \\ \hline x_n & x_2 & \cdots & x_{n-1} & x_n & f \end{array}}.$$

Then, since $F_1 \sim^l D_1 \cap E_n$ we have $f \sim f_1$.

Likewise, exchange the value of f_1 at F_2, that is, x_n, and the value at $D_2 \cap E_n$, that is, x_2, and let

$$f_2 = \frac{\begin{array}{ccccc} F_1 & F_2 & \cdots & F_{n-1} & \\ \hline x_1 & x_2 & \cdots & x_n & \end{array}}{\begin{array}{cccccc} D_1 \cap E_n & D_2 \cap E_n & \cdots & D_{n-1} \cap E_n & D_n \cap E_n & D_n^c \cap E_n^c \\ \hline x_n & x_n & \cdots & x_{n-1} & x_n & f \end{array}}.$$

Then, since $F_2 \sim^l D_2 \cap E_n$we have $f_1 \sim f_2$.

Repeat the same operation, then we obtain

$$f_{n-1} = \begin{array}{cccccc} F_1 & F_2 & \cdots & F_{n-1} & & \\ \hline x_1 & x_2 & \cdots & x_{n-1} & & \\ \hline D_1 \cap E_n & D_2 \cap E_n & \cdots & D_{n-1} \cap E_n & D_n \cap E_n & D_n^c \cap E_n^c \\ \hline x_n & x_n & \cdots & x_n & x_n & f \end{array}$$

and

$$f \sim f_1 \sim f_2 \sim \cdots \sim f_{n-1}.$$

The remaining part is to show $f_{n-1} \sim g$. Now look at

$$f_{n-1} = \begin{array}{cccccc} F_1 & F_2 & \cdots & F_{n-1} & & \\ \hline x_1 & x_2 & \cdots & x_{n-1} & & \\ \hline D_1 \cap E_n & D_2 \cap E_n & \cdots & D_{n-1} \cap E_n & D_n \cap E_n & D_n^c \cap E_n^c \\ \hline x_n & x_n & \cdots & x_n & x_n & f \end{array}$$

$$g = \begin{array}{cccccc} F_1 & F_2 & \cdots & F_{n-1} & & \\ \hline g & g & \cdots & g & & \\ \hline D_1 \cap E_n & D_2 \cap E_n & \cdots & D_{n-1} \cap E_n & D_n \cap E_n & D_n^c \cap E_n^c \\ \hline x_n & x_n & \cdots & x_n & x_n & g \end{array}.$$

Then, by P2, $f_{n-1} \sim g$ is equivalent to $f'_{n-1} \sim g'$, where f'_{n-1}, g' are defined by

$$f'_{n-1} = \begin{array}{cccccc} F_1 & F_2 & \cdots & F_{n-1} & & \\ \hline x_1 & x_2 & \cdots & x_{n-1} & & \\ \hline D_1 \cap E_n & D_2 \cap E_n & \cdots & D_{n-1} \cap E_n & D_n \cap E_n & D_n^c \cap E_n^c \\ \hline x_{n-1} & x_{n-1} & \cdots & x_{n-1} & x_{n-1} & f \end{array}$$

$$g' = \begin{array}{cccccc} F_1 & F_2 & \cdots & F_{n-1} & & \\ \hline g & g & \cdots & g & & \\ \hline D_1 \cap E_n & D_2 \cap E_n & \cdots & D_{n-1} \cap E_n & D_n \cap E_n & D_n^c \cap E_n^c \\ \hline x_{n-1} & x_{n-1} & \cdots & x_{n-1} & x_{n-1} & g \end{array}.$$

Now notice that $f'_{n-1} \sim g'$ follows from the induction assumption. □

Step 4: Define a binary relation $\succsim^*$ over $\Delta_S(X)$ by

$$\Psi \succsim^* \Psi' \iff \Psi = \Psi_{p,f},\ \Psi' = \Psi_{p,f'} \text{ for } f, f' \text{ with } f \succsim f'$$

Note that it follows from convex-rangedness of p that for all $\Psi \in \Delta_S(X)$ there is a simple act f which generates $\Psi = \Psi_{p,f}$.

The remaining part is to show that $\succsim^*$ satisfies axioms of von-Neumann/ Morgenstern expected utility theory. Completeness and transitivity are obvious.

Let us show independence. Given $\Psi, \Psi' \in \Delta_S(X)$, suppose

$$\Psi \succ^* \Psi'.$$

Take simple acts f and g such that $\Psi_{p,f} = \Psi$, $\Psi_{p,g} = \Psi'$, where

$$f = \begin{bmatrix} x_1 & D_1 \\ x_2 & D_2 \\ \vdots & \vdots \\ x_{m-1} & D_{m-1} \\ x_m & D_m \end{bmatrix}, \quad g = \begin{bmatrix} y_1 & E_1 \\ y_2 & E_2 \\ \vdots & \vdots \\ y_{n-1} & E_{n-1} \\ y_n & E_n \end{bmatrix}.$$

Let us see these as

$$f = \begin{array}{c|ccccc} & E_1 & E_2 & \cdots & E_{n-1} & E_n \\ \hline D_1 & x_1 & x_1 & \cdots & x_1 & x_1 \\ D_2 & x_2 & x_2 & \cdots & x_2 & x_2 \\ \vdots & \vdots & \vdots & \cdots & \vdots & \vdots \\ D_{m-1} & x_{m-1} & x_{m-1} & \cdots & x_{m-1} & x_{m-1} \\ D_m & x_m & x_m & \cdots & x_m & x_m \end{array},$$

$$g = \begin{array}{c|ccccc} & E_1 & E_2 & \cdots & E_{n-1} & E_n \\ \hline D_1 & y_1 & y_2 & \cdots & y_{n-1} & y_n \\ D_2 & y_1 & y_2 & \cdots & y_{n-1} & y_n \\ \vdots & \vdots & \vdots & \cdots & \vdots & \vdots \\ D_{m-1} & y_1 & y_2 & \cdots & y_{n-1} & y_n \\ D_m & y_1 & y_2 & \cdots & y_{n-1} & y_n \end{array}.$$

Then, for arbitrary $\lambda \in [0, 1]$, take $\{A_{kl}\}_{k=1,\cdots,m,l=1,\cdots,n}$ such that

$$A_{kl} \subset D_k \cap E_l, \quad p(A_{kl}) = \lambda p(D_k \cap E_l)$$

hold for all $k = 1 \cdots, m$ and $l = 1, \cdots, n$.

Let

$$A = \bigcup_{k=1,\cdots,m} \bigcup_{l=1,\cdots,n} A_{kl},$$

then we have $p(A) = \lambda$.

Now if we could show

$$\begin{bmatrix} f(\omega) & \omega \in A \\ z & \omega \notin A \end{bmatrix} \succ \begin{bmatrix} g(\omega) & \omega \in A \\ z & \omega \notin A \end{bmatrix},$$

then P2 allows us to replace z with act h that yields an arbitrary lottery Ψ'' through $\Psi_{p,h} = \Psi''$, and the proof ends.

From the assumption $f \succ g$, by P6

$$\begin{bmatrix} f(\omega) & \omega \in A \\ z & \omega \notin A \end{bmatrix} \succ \begin{bmatrix} g(\omega) & \omega \in A \\ z & \omega \notin A \end{bmatrix}$$

holds for sufficiently small $\lambda > 0$.

Now for this λ, in the same way we can take $\{B_{kl}\}_{k=1,\cdots,m,l=1,\cdots,n}$ such that $B_{kl} \cap A_{kl} = \emptyset$ and

$$B_{kl} \subset D_k \cap E_l, \quad p(B_{kl}) = \lambda p(D_k \cap E_l)$$

hold for all $k = 1 \cdots, m$ and $l = 1, \cdots, n$. Then $B = \bigcup_{k=1,\cdots,m} \bigcup_{l=1,\cdots,n} B_{kl}$ satisfies $p(B) = \lambda$ and

$$\begin{bmatrix} f(\omega) & \omega \in A \\ z & \omega \notin A \end{bmatrix} \succ \begin{bmatrix} g(\omega) & \omega \in A \\ z & \omega \notin A \end{bmatrix},$$

$$\begin{bmatrix} f(\omega) & \omega \in B \\ z & \omega \notin B \end{bmatrix} \succ \begin{bmatrix} g(\omega) & \omega \in B \\ z & \omega \notin B \end{bmatrix}$$

Hence by applying P2 twice we have

$$\begin{bmatrix} f(\omega) & \omega \in A \\ f(\omega) & \omega \in B \\ z & \omega \notin A \cup B \end{bmatrix} \succ \begin{bmatrix} f(\omega) & \omega \in A \\ g(\omega) & \omega \in B \\ z & \omega \notin A \cup B \end{bmatrix} \succ \begin{bmatrix} g(\omega) & \omega \in A \\ g(\omega) & \omega \in B \\ z & \omega \notin A \cup B \end{bmatrix}$$

and the claim holds for 2λ. Since λ could be taken as arbitrarily small, by repeating this argument we obtain the conclusion for arbitrary $\lambda \in [0, 1]$.

Mixture continuity is left as an exercise.

Since $\succsim^*$ satisfies the axiom of expected utility theory on the set of simple lotteries, there is a function $u : X \to \mathbb{R}$ such that

$$l \succsim^* l' \iff \sum_{z \in S(l)} u(z) l(z) \geq \sum_{z \in S(l')} u(z) l'(z)$$

holds. Thus we obtain

$$\begin{aligned} f \succsim g &\iff \Psi_{p,f} \succsim^* \Psi_{p,g} \\ &\iff \sum_{x \in f(\Omega)} u(x)p(f^{-1}(x)) \geq \sum_{x \in g(\Omega)} u(x)p(g^{-1}(x)). \end{aligned}$$

5.2 Robust Definition of Subjective Belief

P2 has two roles. One is to ensure that subjective belief satisfies additivity. The other is to ensure that the preference over lotteries induced by the subjective belief and acts satisfies the independence axiom. Even when the decision maker's subjective belief is represented by a probability measure, it will not have to be that his preference over lotteries satisfies the independence axiom, however.

Machina and Schmeidler [77] proposed the following axiom, in the form of strengthening of P4, which captures only the first role stated above.

P4* (Strong Comparative Probability): For all disjoint $A, B \subset \Omega$, for all $x^*, x, y^*, y \in X$ with $x^* \succ x$ and $y^* \succ y$, and for all $g, h \in \mathcal{F}$,

$$\begin{bmatrix} x^* & \text{if } \omega \in A \\ x & \text{if } \omega \in B \\ g(\omega) & \text{if } \omega \notin A \cup B \end{bmatrix} \succsim \begin{bmatrix} x & \text{if } \omega \in A \\ x^* & \text{if } \omega \in B \\ g(\omega) & \text{if } \omega \notin A \cup B \end{bmatrix}$$

$$\Longrightarrow \begin{bmatrix} y^* & \text{if } \omega \in A \\ y & \text{if } \omega \in B \\ h(\omega) & \text{if } \omega \notin A \cup B \end{bmatrix} \succsim \begin{bmatrix} y & \text{if } \omega \in A \\ y^* & \text{if } \omega \in B \\ h(\omega) & \text{if } \omega \notin A \cup B \end{bmatrix}.$$

Theorem 5.2 *The following two statements are equivalent:*

(a) $\succsim$ *satisfies P1, 3, 4*, 5, 6.*
(b) There exist a convex-ranged finitely additive probability measure p over Ω and a function $V : \Delta_S(X) \to \mathbb{R}$ such that $\succsim$ is represented by $U : \mathcal{F} \to \mathbb{R}$ in the form

$$U(f) = V(\Psi_{p,f}).$$

Moreover, if another pair (q, W) forms a representation of the same preference, we have $p = q$ and there is a monotone transformation ϕ such that $W = \phi \circ V$.

You can see that the proof of the Savage theorem applies up to Step 3 as it is.

5.3 Anscombe-Aumann Subjective Expected Utility Theory

The weakness of Savage's theory is that it is applicable only when the set of states is "rich." For the Savage theory it is *necessary* that the set of states is essentially a continuum, since any event needs to be partitioned into two subjectively equally likely event.

Hence it does not apply when there are only finite states. Suppose for example that there are only two states, and let $\Omega = \{\omega_1, \omega_2\}$. Then the qualitative probability $\{\omega_1\} \succsim^l \{\omega_2\}$ can imply only that the quantitative probability satisfies $p(\{\omega_1\}) \geq 0.5$ and cannot uniquely pin down subjective belief as a probability measure.

The idea in the subjective expected utility theory by Anscombe and Aumann [6] is to consider that the set of outcomes is "rich" and to borrow information about the decision maker's risk attitude. This is the cost, though, which contrasts with the fact that the Savage theory works with any set of outcomes as far as it has at least two elements.

To illustrate, let us cheat for a while by pretending that we already know the decision maker's vNM index u from his choices over lotteries. Now let p denote his subjective probability that it rains tomorrow. Let x be an outcome that is better than y for him, and consider a bet that gives x if it rains tomorrow and y otherwise. Then subjective expected utility of such a bet is

$$pu(x) + (1 - p)u(y).$$

Now let z denote the certainty equivalent of this bet. Then we obtain an equation

$$pu(x) + (1 - p)u(y) = u(z).$$

By solving the equation for p, we obtain

$$p = \frac{u(z) - u(y)}{u(x) - u(y)}.$$

Since we have used a cheat to pretend that we already know u, below we proceed axiomatically instead.

Let Z be the set of pure outcomes, and let $\Delta_S(Z)$ be the set of simple lotteries over Z.

Let Ω be the set of states of the world. For simplicity, let us assume that Ω is a finite set.

An Anscombe-Aumann act (AA act) is a mapping from Ω into $\Delta_S(Z)$, or an element of the product set $\Delta_S(Z)^\Omega$ in other words. In other words, we consider that the set of outcomes is rich by thinking that the outcomes are lotteries (x, y, z above are already lotteries), not just pure outcomes. Let $\mathcal{H}$ denote the set of all AA acts.

An alternative approach is to consider that the set of pure outcomes Z itself is rich, so that we can take certainty equivalents from the set Z, while it involves technicalities. See for example Wakker [99], Nakamura [83] and Ghirardato et al. [43]. Let's be content with taking lottery outcomes here.

Given AA acts $f, g \in \mathcal{H}$ and $\lambda \in [0, 1]$, we can define a mixture of AA acts, denoted by $\lambda f + (1-\lambda)g \in \mathcal{H}$, by taking

$$(\lambda f + (1-\lambda)g)(\omega) = \lambda f(\omega) + (1-\lambda)g(\omega)$$

for each $\omega \in \Omega$. An interpretation is that at each state ω the decision maker receives $f(\omega)$ with probability λ and $g(\omega)$ with probability $1-\lambda$, while we should keep in mind the assumption of reduction of compound lotteries saying that receiving "$f(\omega)$ with probability λ and $g(\omega)$ with probability $1-\lambda$" is the same as receiving $\lambda f(\omega) + (1-\lambda)g(\omega)$.

We consider the preference relation $\succsim$ defined over $\mathcal{H}$, and impose the following axioms.

Axiom 5.1 (Weak Order) $\succsim$ is complete and transitive.

Axiom 5.2 (Mixture Continuity) For all $f, g, h \in \mathcal{H}$, the sets

$$\{\alpha \in [0, 1] : \alpha f + (1-\alpha)g \succsim h\}$$

and

$$\{\alpha \in [0, 1] : h \succsim \alpha f + (1-\alpha)g\}$$

are closed subsets of $[0, 1]$.

To introduce the third axiom, note that a lottery $l \in \Delta_S(Z)$ is seen as an AA act that yields $l(\omega') = l$ for all $\omega' \in \Omega$. Also, for AA acts $f \in \mathcal{H}$ and $\omega \in \Omega$, the lottery $f(\omega)$ is seen as an AA act that yields $f(\omega)(\omega') = f(\omega)$ for all $\omega' \in \Omega$.

Axiom 5.3 (Monotonicity) For all $f, g \in \mathcal{H}$ if $f(\omega) \succsim g(\omega)$ holds for all $\omega \in \Omega$, then $f \succsim g$.

Monotonicity is meant to ensure state-independence. To understand, consider a "state-dependent" expected utility function

$$U(x_1, x_2) = p_1 u_1(x_1) + p_2 u_2(x_2),$$

while we should note that the probability distribution (p_1, p_2) does not have a well-defined sense of subjective belief, since the form

$$p_1^* u_1^*(x_1) + p_2^* u_2^*(x_2)$$

represents the same ranking as well when $u_1^* = \frac{u_1}{2}$, $u_2^* = \frac{p_2 u_2}{2p_2 - 1}$, $p_1^* = 2p_1$, $p_2^* = 2p_2 - 1$ where $p_1 < \frac{1}{2}$. Having said this, let us verify if Monotonicity is met here.

Consider that the state-dependence allows $a, b, c \in \Delta_S(Z)$ such that

$$U(a,a) = p_1u_1(a) + p_2u_2(a) \geq p_1u_1(c) + p_2u_2(c) = U(c,c), \quad u_1(a) < u_1(c)$$

and

$$U(b,b) = p_1u_1(b) + p_2u_2(b) \geq p_1u_1(c) + p_2u_2(c) = U(c,c), \quad u_2(b) < u_2(c).$$

Then we have

$$U(a,b) = p_1u_1(a) + p_2u_2(b) < p_1u_1(c) + p_2u_2(c) = U(c,c),$$

which is a violation of Monotonicity. What is happening here is that the constant act (a, a) is better than (c, c) since a is sufficiently better than c at State 2, while it is worse at State 1, the constant act (b, b) is better than (c, c) since b is sufficiently better than c at State 1, while it is worse at State 2, and the act (a, b) delivers the worse outcome at each state.

The last axiom is mixture independence.

Axiom 5.4 (Independence) For all $f, g, h \in \mathcal{H}$ and $\lambda \in (0, 1]$, it holds

$$f \succsim g \iff \lambda f + (1-\lambda)h \succsim \lambda g + (1-\lambda)h.$$

An intuition is that if you prefer f over g it has to be that you prefer"f with probability λ and h with probability $1-\lambda$" over "g with probability λ and h with probability $1-\lambda$," since otherwise you change your mind after realization of the probability λ event, which results in a dynamic inconsistency.

There is another "cheat" somehow in the above explanation, since the original interpretation of the mixture operation $\lambda f + (1-\lambda)h$ was that at each state ω the decision maker receives $f(\omega)$ with probability λ and $h(\omega)$ with probability $1-\lambda$, which was already involving a "cheat" of presuming that receiving $\lambda f(\omega) + (1-\lambda)h(\omega)$ is the same as receiving "$f(\omega)$ with probability λ and $h(\omega)$ with probability $1-\lambda$." The additional "cheat" here is to presume that receiving "$f(\omega)$ with probability λ and $h(\omega)$ with probability $1-\lambda$" at each ω is the same as receiving "f with probability λ and h with probability $1-\lambda$," which is assuming that the decision maker is indifferent in order of randomization, that is, whether to randomize before realization of uncertainty or to randomize after realization of uncertainty.

I come back to this issue in the chapter on ambiguity.

Theorem 5.3 *The following two statements are equivalent:*

(a) $\succsim$ *satisfies Weak Order, Mixture Continuity, Monotonicity and Independence.*

(b) *There exist a mixture-linear function $u : \Delta_S(Z) \to \mathbf{R}$ and a probability vector p over Ω such that $\succsim$ is represented in the form*

$$U(h) = \sum_{\omega \in \Omega} u(h(\omega))\ p(\omega).$$

Moreover, u is unique up to positive affine transformations and p is unique.

Proof The theorem is trivial if the decision maker is indifferent between all simple lotteries. Hence we assume that there exist l, l' such that $l \succ l'$.

Step 1: Since $\succsim$ satisfies the axioms of expected utility theory over the set of simple lotteries $\Delta_S(Z)$, it is represented in the form

$$u(l) = \sum_{k=1}^{n} v(x_k) l_k.$$

Without loss of generality, assume we take u such that $u(\Delta_S(Z)) = [\underline{u}, \overline{u}]$ with $\underline{u} < 0 < \overline{u}$.

Step 2: For arbitrary $h \in \mathcal{H}$, define $U(h)$ by

$$U(h) = u(l)$$

for $l \in \Delta_S(Z)$ such that $h \sim l$.

Step 3: Let $\mathcal{U} = \{u \circ h \in \mathbb{R}^\Omega : h \in \mathcal{H}\}$, where $u \circ h \in \mathbb{R}^\Omega$ is a vector (or mapping from Ω to $\mathbb{R}$) defined by $(u \circ h)(\omega) = u(h(\omega))$ for each ω. Then, define a function $I : \mathcal{U} \to \mathbb{R}$ by

$$I(w) = U(h)$$

for $h \in \mathcal{H}$ with $w = u \circ h$. This is well-defined because $u \circ h = u \circ g$ implies $I(u \circ h) = I(u \circ g)$ from Monotonicity.

Step 4: For all $h, h' \in \mathcal{H}$, pick $l, l' \in \Delta_S(Z)$ such that $h \sim l$ and $h' \sim l'$. Then by Independence,

$$\lambda h + (1 - \lambda)h' \sim \lambda l + (1 - \lambda)l'.$$

Hence we obtain

$$\begin{aligned} U(\lambda h + (1 - \lambda)h') &= u(\lambda l + (1 - \lambda)l') \\ &= \lambda u(l) + (1 - \lambda)u(l') \\ &= \lambda U(h) + (1 - \lambda)U(h'). \end{aligned}$$

Therefore, for all $w, w' \in \mathcal{U}$ and $\lambda \in (0, 1]$, by taking $w = u \circ h$ and $w' = u \circ h'$ we show

$$I(\lambda v + (1-\lambda)v') = \lambda I(v) + (1-\lambda)I(v').$$

Since we can take $l \in \Delta_S(Z)$ such that $w' = u(l) = \mathbf{0}$ without loss of generality, for all $v \in \mathcal{U}$ and $\lambda \in (0, 1]$ we obtain

$$I(\lambda w) = \lambda I(w).$$

By dividing the both sides by λ we have

$$I(w) = \frac{1}{\lambda} I(\lambda w).$$

Hence by letting $\lambda' = 1/\lambda$ and $\lambda w = w'$, for all $w' \in \mathcal{U}$ and $\lambda' \geqq 1$ we obtain

$$I(\lambda' w') = \lambda' I(w').$$

Summing up, for all $w \in \mathcal{U}$ and $\lambda > 0$ with $\lambda w \in \mathcal{U}$,

$$I(\lambda w) = \lambda I(w).$$

Thus we can extend I to the entire $\mathbb{R}^{\Omega}$. That is, for all $w \in \mathbb{R}^{\Omega}$, by taking $\lambda > 0$ such that $\lambda w \in \mathcal{U}$ we can set

$$I(w) = \frac{1}{\lambda} I(\lambda w).$$

Now it is easy to show that such extended $I : \mathbb{R}^{\Omega} \to \mathbb{R}$ satisfies

$$I(aw + bw') = aI(w) + bI(w')$$

for all $v, v' \in \mathbb{R}$ and $a, b \geqq 0$.

Step 5: Any monotone and linear function $I : \mathbb{R}^{\Omega} \to \mathbb{R}$ is written in the form

$$I(v) = \sum_{\omega \in \Omega} v(\omega)\ p(\omega)$$

for some $p \in \mathbb{R}^{\Omega}_{+} \setminus \{\mathbf{0}\}$. Without loss, we can set $\sum_{\omega \in \Omega} p(\omega) = 1$. □

Chapter 6
Ambiguity of Belief

6.1 Ellsberg Paradox

In the previous chapter we gave an axiomatic foundation of having subjective belief in the form of a probability measure. It is worth questioning if the decision maker's subjective belief can be described by a single probability measure, an additive set-function. In other words, it is worth questioning the axiom which characterizes a probabilistic belief, that is the eventwise separability axiom or the independence axiom.

The following example is originally due to Ellsberg [28].

Example 6.1 There are 90 balls in the urn, which are red or blue or green. You know that 30 balls are red, but nothing about the numbers of blue and green.

Consider the following four bets.

	R	B	G
f_1	\$100	\$0	\$0
f_2	\$0	\$100	\$0
f_3	\$100	\$0	\$100
f_4	\$0	\$100	\$100

Verbally, they say

f_1 gives you 100 if red is drawn, and nothing otherwise.
f_2 gives you 100 if blue is drawn, and nothing otherwise.
f_3 gives you 100 if red or green is drawn, and nothing otherwise.
f_4 gives you 100 if red or green is drawn, and nothing otherwise (i.e., if red is drawn).

T. Hayashi, *Decision Theory*, Monographs in Mathematical Economics 9,
https://doi.org/10.1007/978-981-95-2200-2_6

First compare between f_1 and f_2. Then, since the probability of red is known while that of blue is unknown, it is natural to conclude that

$$f_1 \succ f_2.$$

Also, compare between f_3 and f_4 Then, since the probability of even "red or green" is known, while that of "blue or green" is unknown, it is natural to conclude that

$$f_3 \prec f_4.$$

But these two choices contradict the subjective expected utility theory. Let us see this first at the level of representation. The first choice is represented by

$$p(R)u(100) + p(B \cup G)u(0) > p(B)u(100) + p(R \cup G)u(0).$$

Then by additivity of probability measure we have $p(B \cup G) = p(B) + p(G)$ and $p(R \cup G) = p(R) + p(G)$, we obtain (by assuming $u(100) > u(0)$)

$$p(R) > p(B).$$

On the other hand, the second choice is represented by

$$p(R \cup G)u(100) + p(B)u(0) < p(B \cup G)u(100) + p(R)u(0).$$

Again by additivity of probability measure, we obtain

$$p(R) < p(B),$$

which is a contradiction to the previous implication.

At an axiomatic level this is a direct contradiction to P2, Eventwise Separability, for

$$f_3 = \begin{bmatrix} f_1 & \text{if } R \cup B \\ 100 & \text{if } G \end{bmatrix}, \quad f_4 = \begin{bmatrix} f_2 & \text{if } R \cup B \\ 100 & \text{if } G \end{bmatrix}.$$

In the Anscombe-Aumann setting, this means that the independence axiom is violated (although it is not violated regarding lotteries, i.e., constant acts). It will be exemplified as follows.

	R	B	G
h_1	\$100	\$0	\$0
h_2	\$0	\$100	\$0
h_3	(\$ 100; 0.5, \$ 0; 0.5)	\$0	(\$100; 0.5, \$0; 0.5)
h_4	\$0	(\$100; 0.5, \$0; 0.5)	(\$100; 0.5, \$0; 0.5)

By the same reasoning we will observe $h_1 \succ h_2$ and $h_3 \prec h_4$.

Let h_5 denote an AA-act that gives 100 for sure if green is drawn and 0 for sure otherwise. Then, since $h_3 = 0.5h_1 + 0.5h_5$ and $h_4 = 0.5h_2 + 0.5h_5$ we obtain a contradiction to Independence.

The tendency to prefer a bet on an event with known probability over a bet on an event with unknown probability is called **ambiguity aversion**. The Ellsberg paradox is an example of ambiguity aversion. Below, we present some axiomatic theories of decision criteria that allow ambiguity aversion.

6.2 Maxmin Expected Utility with Multiple Priors

First, let us illustrate the maxmin expected utility model with multiple priors, which represents the decision maker's belief as a set of probability distributions instead of a single probability distribution, that is due to Gilboa and Schmeidler [45].

We work on the Anscombe-Aumann domain, in which outcomes are taken to be lotteries. For approaches not using lottery outcomes, see Casadesus-Masanell et al. [14], Ghirardato et al. [43], Alon and Schmeidler [5] for example.

Let Z be the set of pure outcomes and let $\Delta_S(Z)$ be the set of simple lotteries over Z. Let Ω be the set of states, which is assumed to be finite for simplicity. An Anscombe-Aumann act, or an AA act, is a mapping from Ω into $\Delta_S(Z)$, or an element of the product set $\Delta_S(Z)^\Omega$. Let $\mathcal{H} = \Delta_S(Z)^\Omega$ be the set of such acts.

Given AA acts $f, g \in \mathcal{H}$ and $\lambda \in [0, 1]$, define a mixture of AA acts $\lambda f + (1 - \lambda)g \in \mathcal{H}$ by letting

$$(\lambda f + (1 - \lambda)g)(\omega) = \lambda f(\omega) + (1 - \lambda)g(\omega)$$

for each $\omega \in \Omega$.

Let $\succsim$ denote the preference ranking over $\mathcal{H}$. We impose the following axioms.

Axiom 6.1 (Weak Order) $\succsim$ satisfies completeness and transitivity.

Axiom 6.2 (Mixture Continuity) For all $f, g, h \in \mathcal{H}$, the sets

$$\{\alpha \in [0, 1] : \alpha f + (1 - \alpha)g \succsim h\}$$

and

$$\{\alpha \in [0, 1] : h \succsim \alpha f + (1 - \alpha)g\}$$

are closed subsets of [0, 1].

Axiom 6.3 (Monotonicity) For all $f, g \in \mathcal{H}$ if $f(\omega) \succsim g(\omega)$ holds for all $\omega \in \Omega$, then $f \succsim g$.

Axiom 6.4 (Certainty Independence) For all $f, g \in \mathcal{H}$, for all $l \in \Delta_S(Z)$ and $\lambda \in (0, 1]$,

$$f \succsim g \iff \lambda f + (1 - \lambda)l \succsim \lambda g + (1 - \lambda)l.$$

Note that here a lottery l is taken to be a constant act (function) which yields $l(\omega') = l$ for all $\omega' \in \Omega$. Hereafter we refer to constants and constant functions in an interchangeable manner.

Certainty Independence states that a mixture with a constant act does not change how payoffs are distributed across states in the relative sense therefore ambiguity aversion has no role to play.

Axiom 6.5 (Uncertainty Aversion) For all $f, g \in \mathcal{H}$ and $\lambda \in [0, 1]$,

$$f \sim g \implies \lambda f + (1 - \lambda)g \succsim f.$$

This means that taking a mixture of AA acts smoothes utilities/payoffs across states and allows the decision maker to avoid uncertainty. For example, suppose f gives 100 dollars for sure at State 1 and 0 dollars for sure at State 2, g gives 0 dollars for sure at State 1 and 100 dollars for sure at State 2, then $\frac{f+g}{2}$ gives an even-chances lottery over 100 and 0 at both states.

Theorem 6.1 *The following two statements are equivalent:*

(a) $\succsim$ *satisfies Weak Order, Mixture Continuity, Monotonicity, Certainty Independence and Uncertainty Aversion.*
(b) There exists a mixture-linear function $u : \Delta_S(X) \to \mathbf{R}$ *and a closed convex set* P *of probability vectors over* Ω *such that* $\succsim$ *is represented in the form*

$$U(h) = \min_{p \in P} \sum_{\omega \in \Omega} u(h(\omega))\ p(\omega).$$

Moreover, u *is unique up to positive affine transformations and* P *is unique.*

Proof The proof is the same as the one for the Anscombe-Aumann subjective expected utility representation theorem, up to Step 3. Thus we start from Step 4.

Step 4: For an arbitrary $h \in \mathcal{H}$, pick any $l \in \Delta_S(Z)$ with $h \sim l$. Then, for all $l' \in \Delta_S(Z)$, by Certainty Independence,

$$\lambda h + (1-\lambda)l' \sim \lambda l + (1-\lambda)l'.$$

Hence we obtain

$$\begin{aligned} U(\lambda h + (1-\lambda)l') &= u(\lambda l + (1-\lambda)l') \\ &= \lambda u(l) + (1-\lambda)u(l'). \end{aligned}$$

Thus, for all $v \in \mathcal{U}$, for all $c \in \mathbb{R}$ with $c\mathbf{1} \in \mathcal{U}$, and for all $\lambda \in (0, 1]$, by taking h and l such that $v = u \circ h$ and $c\mathbf{1} = u \circ l$, we obtain

$$I(\lambda v + (1-\lambda)c\mathbf{1}) = \lambda I(v) + (1-\lambda)c.$$

Note that $\mathbf{1}$ denotes the vector of ones.

By applying the above to the case of $c = u(l^*) = 0$, we obtain

$$I(\lambda v) = \lambda I(v)$$

for all $v \in \mathcal{U}$ and $\lambda \in (0, 1]$. By dividing both sides by λ, we obtain

$$I(v) = \frac{1}{\lambda} I(\lambda v).$$

Hence by letting $\lambda' = 1/\lambda$ and $\lambda v = v'$, we obtain

$$I(\lambda' v') = \lambda' I(v')$$

for arbitrary $v' \in \mathcal{U}$ and $\lambda' \geqq 1$. Summing up, for all $v \in \mathcal{U}\, \lambda > 0$ with $\lambda v \in \mathcal{U}$,

$$I(\lambda v) = \lambda I(v).$$

From this we can extend I to the function defined over the entire $\mathbb{R}^\Omega$. That is, for arbitrary $v \in \mathbb{R}^\Omega$, by taking $\lambda > 0$ such that $\lambda v \in \mathcal{U}$ we can define

$$I(v) = \frac{1}{\lambda} I(\lambda v).$$

Now we can show that the extended function $I : \mathbb{R}^\Omega \to \mathbb{R}$ satisfies

$$I(av + b\mathbf{1}) = aI(v) + b$$

for arbitrary $v \in \mathbb{R}^\Omega$ and $a, b \geqq 0$.

Step 5: Without loss, focus on the set

$$V = \{v \in \mathbb{R}^{\Omega} : I(v) \geqq 0\}.$$

Then V is a closed convex cone pointing to $\mathbf{0}$. It follows from Monotonicity that $V \supset \mathbb{R}^{\Omega}_{+}$. By the supporting hyperplane theorem, there is a vector $p \in \mathbb{R}^{\Omega} \setminus \{\mathbf{0}\}$ such that

$$p \cdot \mathbf{0} \leqq p \cdot v$$

holds for all $v \in V$. Since $V \supset \mathbb{R}^{\Omega}_{+}$ we have $p \in \mathbb{R}^{\Omega}_{+}$. For, suppose with $p(\omega) < 0$ for some $\omega \in \Omega$, then $p \cdot e_{\omega} = p(\omega) < 0$, where e_{ω} denotes the vector that takes 1 on the ω-th coordinate and 0 elsewhere, which contradicts the supporting hyperplane condition. Since p is nonzero, we can take $\sum_{\omega \in \Omega} p(\omega) = 1$ without loss.

Denote the set consisting of such p by P, then it is clearly a convex set. Also, since V is closed P is closed as well. Hence P is a compact subset of $\mathbb{R}^{\Omega}$.

Now, for an arbitrary $v \in \mathbb{R}^{\Omega}$, if $I(v) \geqq 0$ then $v \in V$ by the definition of V. Hence $p \cdot v \geqq 0$ holds for all $p \in P$, which implies $\min_{p \in P} p \cdot v \geqq 0$.

Moreover, if $\min_{p \in P} p \cdot v < 0$, then there is $p \in P$ such that $p \cdot v < 0$. Note that by the definition of P, $p \cdot v' \geq 0$ holds for all $v' \in V$, and it must be that $v \notin V$. Hence $I(v) < 0$.

Thus we obtain

$$I(v) \geqq 0 \iff \min_{p \in P} p \cdot v \geqq 0.$$

Finally, for general $v \in \mathbb{R}^{\Omega}$ with $I(v) = c$ we can take $v' \in V$ such that $I(v') = 0$ and $v = v' + c\mathbf{1}$. Thus we have

$$\begin{aligned} I(v) &= I(v' + c\mathbf{1}) \\ &= I(v') + c \\ &= \min_{p \in P} p \cdot v' + c \\ &= \min_{p \in P} p \cdot (v' + c\mathbf{1}) \\ &= \min_{p \in P} p \cdot v. \end{aligned}$$

To show the uniqueness of P, suppose there is a closed convex set $Q \neq P$ such that

$$I(v) = \min_{p \in Q} p \cdot v$$

holds for all $v \in \mathbb{R}^{\Omega}$.

From $Q \neq P$ without loss we can think of a probability vector p^* with $p^* \in P$ and $p^* \notin Q$. Then by the separating hyperplane theorem there is a vector $v \in \mathbb{R}^\Omega \setminus \{\mathbf{0}\}$ and a number α such that

$$p^* \cdot v < \alpha \leqq q \cdot v$$

holds for all $q \in Q$.

Hence we have $I(v) = \min_{q \in Q} q \cdot v \geqq \alpha$. On the other hand, we have $I(v) = \min_{p \in P} p \cdot v \leqq p^* \cdot v < \alpha$ and thus obtain a contradiction.

Uniqueness of u is standard. □

In the leading Ellsberg paradox example, an extreme but simple resolution is to set

$$P = \{(p(R), p(B), p(G)) : p(R) = 1/3\}.$$

Then, in choosing f_2 the worst case is $p(B) = 0$, which yields

$$U(f_1) = \frac{1}{3}u(100) + \frac{2}{3}u(0) > u(0) = U(f_2).$$

Also, in choosing f_3 the worst case is $p(G) = 0$, which yields

$$U(f_3) = \frac{1}{3}u(100) + \frac{2}{3}u(0) < \frac{2}{3}u(100) + \frac{1}{3}u(0) = U(f_4).$$

Thus it resolves the paradox.

6.3 Non-additive Subjective Expected Utility

Next, let us present the model of non-additive subjective expected utility due to Schmeidler [92], in which the decision maker's belief is described as a general set function, which may not be additive, instead of a probability measure that is an additive set function.

Again, we work on the Anscombe-Aumann domain in which outcomes are taken to be lotteries. For approaches not using lottery outcomes, see for example Nakamura [83], Ghirardato et al. [43], who consider that the set of pure outcomes is rich, and Sarin and Wakker [90] who consider that the set of states is rich as in Savage's theory.

Consider preference $\succsim$ defined over the set of Anscombe-Aumann acts $\mathcal{H}$.

Definition 6.1 Two acts $f, g \in \mathcal{H}$ are said to be **comonotonic** if

$$f(\omega) \succsim f(\omega') \iff g(\omega) \succsim g(\omega')$$

holds for all $\omega, \omega' \in \Omega$.

Verbally, two bets are said to be comonotonic if they take the same *position*. For example, when f gives you 100 if red, 50 if blue and 70 if green, and when g gives you 80 if red, 20 if blue and 30 if green, they are comonotonic since red is the first-best, green is the second-best and blue is the worst case for both.

Axiom 6.6 (Comonotonic Independence) For all $f, g, h \in \mathcal{H}$ which are comonotonic to each other and for all $\lambda \in (0, 1]$,

$$f \succsim g \iff \lambda f + (1-\lambda)h \succsim \lambda g + (1-\lambda)h.$$

Comonotonic Independence states that ambiguity aversion has no role to play when the mixture does not change the position of acts. Since constant acts are comonotonic to any act, Comonotonic Independence implies Certainty Independence. To see this, for any $f, g \in \mathcal{H}$, take $l, l' \in \Delta_S(Z)$ such that $f \sim l$ and $g \sim l'$ respectively. Pick any $l'' \in \Delta_S(Z)$, then since f, l, l'' are comonotonic, $\lambda f + (1-\lambda)l'' \sim \lambda l + (1-\lambda)l''$. Also, since g, l', l'' are comonotonic we have $\lambda g + (1-\lambda)l'' \sim \lambda l' + (1-\lambda)l''$. Thus we obtain equivalence between $f \succsim g$ and $\lambda f + (1-\lambda)l'' \succsim \lambda g + (1-\lambda)l''$.

Theorem 6.2 *The two statements below are equivalent:*

(a) $\succsim$ *satisfies Weak Order, Mixture Continuity, Monotonicity and Comonotonic Independence.*

(b) There exists a mixture-linear function $u : \Delta_S(X) \to \mathbf{R}$ *and a monotone set function* $p : 2^{\Omega} \to [0, 1]$ *with* $p(\emptyset) = 0$ *and* $p(\Omega) = 1$ *such that preference* $\succsim$ *is represented in the form*

$$U(h) = \sum_{k=1}^{|h(\Omega)|} u(x_k)\left[p\left(\bigcup_{j=1}^{k} E_j\right) - p\left(\bigcup_{j=1}^{k-1} E_j\right)\right]$$

for all $h = (x_1; E_1, x_2; E_2, \cdots, x_n; E_n)$ *that is suitably rewritten so that* $u(x_1) \geq u(x_2) \geq \cdots \geq u(x_n)$.

Moreover, u is unique up to positive affine transformations and p is unique.

Proof The proof is the same as the one for the Anscombe-Aumann subjective expected utility representation theorem, up to Step 3. Thus we start from Step 4.

Step 4: Because Comonotonic Independence implies Certainty Independence, Step 4 in the proof of Theorem 6.1 goes through as it is.

Now we can show from Comonotonic Independence that the extended function $I : \mathbb{R}^\Omega \to \mathbb{R}$ satisfies

$$I(av + bw) = aI(v) + bI(w)$$

for any two vectors $v, w \in \mathbb{R}^\Omega$ which are comonotonic to each other and $a, b \geqq 0$.

Step 5: Define a set function $p : 2^\Omega \to [0, 1]$ by

$$p(E) = I(\mathbf{1}_E)$$

for each $E \in 2^\Omega$, where $\mathbf{1}_E$ denotes the vector such that its coordinate is 1 on E and 0 elsewhere. Then it satisfies $p(\emptyset) = 0$ and $p(\Omega) = 1$.

Now we show that for all $n \geqq 1$, for all n numbers $a_1 > a_2 > \cdots > a_n > 0$ with descending order, for all $E_1, E_2, \cdots, E_{n-1}, E_n$ being an n-partition of Ω the vector $a = \sum_{k=1}^n a_k \mathbf{1}_{E_k}$ satisfies

$$I(a) = \sum_{k=1}^{n} a_k \left[p\left(\bigcup_{j=1}^{k} E_j\right) - p\left(\bigcup_{j=1}^{k-1} E_j\right) \right].$$

We prove this by induction.

When $n = 1$, since I is homogeneous of degree 1 and from the definition of p, we have

$$I(a_1 \mathbf{1}_{E_1}) = a_1 I(\mathbf{1}_{E_1}) = a_1 p(E_1).$$

For general n, note that vector $a = \sum_{k=1}^n a_k \mathbf{1}_{E_k}$ is written as $a = b + c$, where $b = \sum_{k=1}^{n-1} (a_k - a_n) \mathbf{1}_{E_k}$ and $c = a_n \mathbf{1}_{\bigcup_{k=1}^n E_k}$.

From the induction hypothesis,

$$I(b) = \sum_{k=1}^{n-1} (a_k - a_n) \left[p\left(\bigcup_{j=1}^{k} E_j\right) - p\left(\bigcup_{j=1}^{k-1} E_j\right) \right],$$

$$I(c) = a_n p\left(\bigcup_{k=1}^{n} E_k\right).$$

Since b and c are comonotonic, from additivity over comonotonic vectors we obtain

$$\begin{aligned} I(a) &= I(b + c) \\ &= I(b) + I(c) \end{aligned}$$

$$= \sum_{k=1}^{n-1}(a_k - a_n)\left[p\left(\bigcup_{j=1}^{k}E_j\right) - p\left(\bigcup_{j=1}^{k-1}E_j\right)\right] + a_n p\left(\bigcup_{k=1}^{n}E_k\right)$$
$$= \sum_{k=1}^{n} a_k\left[p\left(\bigcup_{j=1}^{k}E_j\right) - p\left(\bigcup_{j=1}^{k-1}E_j\right)\right].$$

□

In the leading Ellsberg paradox example, an extreme but simple resolution is to take

$$p(B) = p(G) = 0,\ \ p(R) = 1/3,\ \ p(R \cup B) = p(R \cup G) = 1/3,\ \ p(B \cup G) = 2/3.$$

Then we have

$$U(f_1) = \frac{1}{3}u(100) + \frac{2}{3}u(0) > u(0) = U(f_2)$$

and

$$U(f_3) = \frac{1}{3}u(100) + \frac{2}{3}u(0) < \frac{2}{3}u(100) + \frac{1}{3}u(0) = U(f_4),$$

which resolves the paradox.

6.4 Comparative Ambiguity Aversion

To see that ambiguity aversion is distinct from risk aversion, let us introduce the following definition (Ghirardato and Marinacci [42]).

Definition 6.2 Say that $\succsim_1$ is more ambiguity averse than $\succsim_2$ if

$$l \succsim_2 h \quad \Longrightarrow \quad l \succsim_1 h$$

and

$$l \succ_2 h \quad \Longrightarrow \quad l \succ_1 h$$

hold for all $l \in \Delta(Z)$ and $h \in \mathcal{H}$.

It means, when $\succsim_2$ prefers a bet with known probability and dislikes a bet with unknown probability, that should be even more the case for $\succsim_1$. Or, by taking the contrapositive, it says when $\succsim_1$ dares to choose a bet with unknown probability, that should be even more the case for $\succsim_2$.

Note that $\succsim_1$ and $\succsim_2$ coincide over lotteries. Thus, the comparative ambiguity aversion ranking is an incomplete one in the sense that we can compare ambiguity attitudes only when risk preferences are identical.

Theorem 6.3 *Suppose $\succsim_1$ and $\succsim_2$ satisfy the axioms for maxmin expected utility and they are represented with let (u_1, P_1) and (u_2, P_2), respectively. Then, $\succsim_1$ is more ambiguity-averse than $\succsim_2$ if and only if there are constants $\alpha > 0$ and β such that*

$$u_1 = \alpha u_2 + \beta$$

and

$$P_1 \supset P_2.$$

Proof The "if" part is straightforward, so we show the "only if" part.

By restricting attention to lotteries, $\succsim_1$ and $\succsim_2$ are identical risk preferences. Hence u_1 and u_2 are identical up to a positive affine transformation.

Now suppose $P_1 \not\supset P_2$, that is, there is some $p \in P_2$ with $p \notin P_1$. By the separating hyperplane theorem, there is a non-zero vector x and a number α such that

$$\langle q, x \rangle \geq \alpha > \langle p, x \rangle \quad \forall q \in P_1.$$

Without loss of generality, we can take $h \in \mathcal{H}$ and $l \in \Delta(Z)$ such that $x = u \circ h$ and $\alpha = u(l)$. Then

$$\langle q, u \circ h \rangle \geq u(l) > \langle p, u \circ h \rangle \quad \forall q \in P_1,$$

which implies

$$\min_{q \in P_1} \langle q, u \circ h \rangle \geq u(l) > \langle p, u \circ h \rangle \geq \min_{q \in P_2} \langle q, u \circ h \rangle.$$

Thus we see $h \succsim_1 l$ despite $l \succ_2 h$, which is a contradiction. □

6.5 Second-Order Prior and Ambiguity Aversion

If you don't know the true probability distribution, why not have a probabilistic belief over probability distributions, and apply another level of risk aversion regarding uncertainty about true probability distribution, which should be understood as observed ambiguity aversion? This would be a natural response.

In other words, we may see the set of probability distributions over states of the world as a larger set of "states" and apply subjective expected utility theory.

Let

$$U(f, p) = \int_{\Omega} u(f(\omega))p(d\omega)$$

be the expected utility representation with given probability distribution p, where u explains risk aversion under given distribution. Then, a second-order prior is given as an element $\mu \in \Delta(\Delta(\Omega))$, and a function ϕ is supposed to explain risk aversion regarding uncertainty about true probability distribution, and we can think of a functional form

$$V(f) = \int_{\Delta(\Omega)} \phi\,(U(f, p))\,\mu(dp).$$

How can we identify this second-order prior μ and the function ϕ from observed choice? Recall that a choice object in subjective expected utility theory is a bet on an event such as "win 100 dollars if it rains tomorrow." If we were to see the set of probability distributions as the set of states, we have to think of a bet such as "win 100 dollars if this type of probability distribution is correct." We can verify if it rained or not ex-post, hence a bet on weather can be an object of choice. But we cannot verify if a particular probability distribution was "correct" or not, unless we introduce a certain form of information at a different level (we come to such type of information explicitly in the next section).

There are two approaches to deal with this issue. One is to think of such bets indeed, at least hypothetically. This approach has been taken by Klibanoff et al. [67]. The other is to think of a richer choice domain, which consists of lotteries over lottery-acts, in which attitude toward ambiguity is revealed through a non-indifference to timing of randomization. This approach is taken by Seo [94].

Let us go over the two approaches in a simplified manner.

6.5.1 The Second-Order Act Approach

To illustrate the first approach, let Ω be the set of states, which is assumed to be finite. We consider Anscombe-Aumann acts. Let Z be the set of pure outcomes, which is assumed to be finite. An act is a mapping from Ω into $\Delta(Z)$. Let $\mathcal{H} = \Delta^{\Omega}$ be the set of AA acts.

We also consider second-order acts. Since $\Delta(\Omega)$ and $\Delta(Z)$ are subsets of Euclidian spaces, they are given σ-algebras of Lebesgue measurable sets. Let $\widetilde{\mathcal{H}}$ be the set of Lebesgue-measurable mappings from $\Delta(\Omega)$ to $\Delta(Z)$. In another words, $\widetilde{\mathcal{H}}$ is the domain of Savage acts in which the state space is $\Delta(\Omega)$ and the set of outcomes is $\Delta(Z)$.

Consider a pair of preference rankings, $(\succsim, \succsim^2)$, where $\succsim$ is defined over $\mathcal{H}$ and $\succsim^2$ is defined over $\widetilde{\mathcal{H}}$.

First, we assume that the preference $\succsim$ satisfies the expected utility theory due to von-Neumann and Morgenstern, only over $\Delta(Z)$. This is justified by adopting the axioms as illustrated in Chap. 4.

Axiom 6.7 (EU over Lotteries) There is a mixture-linear function $u : \Delta(Z) \to \mathbb{R}$ which represents $\succsim$ over $\Delta(Z)$.

We assume that the second-order preference over second-order acts follows the subjective expected utility theory, in which the second-order acts are not necessarily simple and subjective belief is not just finitely additive. We need a justification beyond what we gave in Chap. 5. This can be done by following Arrow [8].

Axiom 6.8 (SEU over Second-Order Acts) There is a countably additive probability measure μ over $\Delta(\Omega)$ and a function $v : \Delta(Z) \to \mathbb{R}$ such that $\succsim^2$ is represented in the form

$$U^2(\widetilde{h}) = \int_{\Delta(\Omega)} v(\widetilde{h}(p))\mu(dp).$$

Lastly, we impose the condition of consistency between the primary preference and the second-order preference. Given $h \in \mathcal{H}$, its second-order counterpart $h^2 \in \widetilde{\mathcal{H}}$ is defined by

$$h^2(p) = \sum_{\omega \in \Omega} h(\omega)p(\omega)$$

for each $p \in \Delta(\Omega)$. Note that the second-order act induced by a lottery (constant act) $l \in \Delta(Z)$ is taken to be l itself.

Axiom 6.9 (Consistency) For all $f, g \in \mathcal{H}$,

$$f \succsim g \iff f^2 \succsim^2 g^2.$$

Theorem 6.4 *A pair* $(\succsim, \succsim^2)$ *satisfies EU over lotteries, SEU over second-order acts and Consistency if and only if there exist a countably additive probability measure* μ *over* $\Delta(\Omega)$*, a mixture-linear function* $u : \Delta(Z) \to \mathbb{R}$ *and an increasing function* $\phi : u(\Delta(Z)) \to \mathbb{R}$ *such that* $\succsim$ *is represented in the form*

$$V(f) = \int_{\Delta(\Omega)} \phi\left(\sum_{\omega \in \Omega} u(f(\omega))p(\omega)\right) \mu(dp)$$

and $\succsim^2$ *is represented in the SEU form.*

Proof Necessity of the axioms is routine. We prove sufficiency.

Since $\succsim$ satisfies the expected utility theory over $\Delta(Z)$, it is represented by a mixture-linear function $u : \Delta(Z) \to \mathbb{R}$.

Let (v, μ) be the pair given under SEU over second-order acts.

By Consistency, $\succsim$ and $\succsim^2$ coincide over $\Delta(Z)$. Hence v and u are *ordinally* equivalent. Hence there is an increasing function ϕ such that

$$v(l) = \phi(u(l))$$

holds for all $l \in \Delta(Z)$.

Then, from Consistency and mixture-linearity of u we have

$$\begin{aligned}
& f \succsim g \\
\Longleftrightarrow\ & f^2 \succsim^2 g^2 \\
\Longleftrightarrow\ & \int_{\Delta(\Omega)} v(f^2(p))\mu(dp) \geq \int_{\Delta(\Omega)} v(g^2(p))\mu(dp) \\
\Longleftrightarrow\ & \int_{\Delta(\Omega)} \phi(u(f^2(p)))\mu(dp) \geq \int_{\Delta(\Omega)} \phi(u(g^2(p)))\mu(dp) \\
\Longleftrightarrow\ & \int_{\Delta(\Omega)} \phi\left(\sum_{\omega\in\Omega} u(f(\omega))p(\omega)\right)\mu(dp) \geq \int_{\Delta(\Omega)} \phi\left(\sum_{\omega\in\Omega} u(g(\omega))p(\omega)\right)\mu(dp).
\end{aligned}$$

□

6.5.2 *The Randomization-over-Acts Approach*

Now we illustrate the second approach. We assume the set of pure outcomes Z is finite. Hence the set of (one-stage) lotteries $\Delta(Z)$ is a compact convex subset of an Euclidian space. Let Ω be the set of states, which is assumed to be finite. Then the set of Anscombe-Aumann acts $\mathcal{H} = \Delta(Z)^{\Omega}$ is a compact and convex subset with respect to the product topology. Now, let $\Delta(\mathcal{H})$ be the set of Borel probability measures over $\mathcal{H}$, which is a compact metric space with respect to the Prokhorov metric.

Consider preference ranking $\succsim$ defined over $\Delta(\mathcal{H})$. The first axiom will need no explanation in the current context.

Axiom 6.10 (Continuous Weak Order) $\succsim$ is a complete, transitive and continuous binary relation over $\Delta(\mathcal{H})$.

To emphasize the difference in layers of mixtures, we denote the mixture operation over $\Delta(Z)$ by $\lambda l \oplus (1-\lambda)m$ for $l, m \in \Delta(Z)$ and $\lambda \in (0, 1)$.

Axiom 6.11 (Second-Stage Independence) For all $l, m, n \in \Delta(Z)$ and for all $\lambda \in (0, 1)$,

$$l \succsim m \text{ if and only if } \lambda l \oplus (1-\lambda)n \succsim \lambda m \oplus (1-\lambda)n.$$

Axiom 6.12 (First-Stage Independence) For all $L, M, N \in \Delta(\mathcal{H})$ and for all $\lambda \in (0, 1)$,

$$L \succsim M \text{ if and only if } \lambda L + (1-\lambda)N \succsim \lambda M + (1-\lambda)N.$$

Given $f \in \mathcal{H}$ and $p \in \Delta(\Omega)$, one can define a **one-stage** lottery $\Psi(f, p) \in \Delta(Z)$ by

$$\Psi(f, p) = \bigoplus_{\omega \in \Omega} f(\omega)p(\omega).$$

Given $L \in \Delta(\mathcal{H})$ and $p \in \Delta(\Omega)$, one can define a **two-stage** lottery $\Psi(L, p) \in \Delta(\Delta(Z))$ by

$$\Psi(L, p)(B) = L(\{f \in \mathcal{H} : \Psi(l, p) \in B\}),$$

where B is an arbitrary Borel subset of $\Delta(Z)$.

Now we state the final axiom.

Axiom 6.13 (Dominance) For all $L, M \in \Delta(\mathcal{H})$ and for all $p \in \Delta(\Omega)$,

$$\Psi(L, p) \succsim \Psi(M, p) \ \forall p \in \Delta(\Omega) \implies L \succsim M.$$

Theorem 6.5 *$\succsim$ satisfies Continuous Weak Order, First-Stage Independence, Second-Stage Independence and Dominance if and only if there is a mixture-linear function $u : \Delta(Z) \to \mathbb{R}$, a regular Borel probability measure μ over $\Delta(\Omega)$ and an increasing function $\phi : u(\Delta(X)) \to \mathbb{R}$ such that $\succsim$ is represented in the form*

$$V(L) = \int_{\mathcal{H}} U(f)dL(f),$$

$$U(f) = \int_{\Delta(\Omega)} \phi\left(\sum_{\omega \in \Omega} u(f(\omega))p(\omega)\right) d\mu(p).$$

Proof From Continuous Weak Order and Second-Stage Independence, $\succsim$ restricted to $\Delta(Z)$ is represented by a mixture-linear function $u : \Delta(Z) \to \mathbb{R}$.

From Continuous Weak Order and First-Stage Independence, there is a continuous function $U : \mathcal{H} \to \mathbb{R}$ such that $\succsim$ is represented in the form

$$V(L) = \int_{\mathcal{H}} U(f)dL(f).$$

Then we have

$$V(\Psi(L,p)) = \int_{\mathcal{H}} U(\Psi(f,p))dL(f)$$

and this is continuous in p.

By Dominance,

$$\int_{\mathcal{H}} U(\Psi(f,p))dL(f) \geq \int_{\mathcal{H}} U(\Psi(f,p))dM(f) \ \forall p \in \Delta(\Omega)$$
$$\Longrightarrow \int_{\mathcal{H}} U(f)dL(f) \geq \int_{\mathcal{H}} U(f)dM(f)$$

holds for all $L, M \in \Delta(\mathcal{H})$.

Let

$$Z = \left\{ \left(\int_{\mathcal{H}} U(f)dL(f) - \int_{\mathcal{H}} U(f)dM(f), \left(\int_{\mathcal{H}} U(\Psi(f,p))dL(f) - \int_{\mathcal{H}} U(\Psi(f,p))dM(f) \right) \right) \in \mathbb{R} \times C(\Delta(\Omega)) : L, M \in \Delta(\mathcal{H}) \right\}.$$

Then it is a convex subset of $\mathbb{R} \times C(\Delta(\Omega))$.

By Dominance,

$$Z \cap ((-\mathbb{R}_{++}) \times C_{+}(\Delta(\Omega))) = \emptyset,$$

where $C_{+}(\Delta(\Omega))$ denotes the set of non-negative continuous functions over $\Delta(\Omega)$.

By the separating hyperplane theorem applied to the space $\mathbb{R} \times C(\Delta(\Omega))$, where $\Delta(\Omega)$ is a compact Hausdorff space, there is a non-zero $(\alpha, -\mu) \in \mathbb{R} \times \mathcal{M}(\Delta(\Omega))$ such that

$$\alpha \left(\int_{\mathcal{H}} U(f)dL(f) - \int_{\mathcal{H}} U(f)dM(f) \right) - \int_{\Delta(\Omega)} \left(\int_{\mathcal{H}} U(\Psi(f,p))dL(f) - \int_{\mathcal{H}} U(\Psi(f,p))dM(f) \right) \mu(dp) \geq \gamma$$

for all $L, M \in \Delta(\mathcal{H})$ and

$$\gamma \geq \alpha u - \int_{\Delta(\Omega)} v(p)d\mu(p)$$

for all $(u, v) \in (-\mathbb{R}_{++}) \times C_{+}(\Delta(\Omega))$. Note that $\mathcal{M}(\Delta(\Omega))$ denotes the set of signed regular Borel measures over $\Delta(\Omega)$.

Since $z \in Z$ implies $-z \in Z$, we have $\gamma = 0$ and

$$\alpha\left(\int_{\mathcal{H}} U(f)dL(f) - \int_{\mathcal{H}} U(f)dM(f)\right) - \int_{\Delta(\Omega)}\left(\int_{\mathcal{H}} U(\Psi(f,p))dL(f) - \int_{\mathcal{H}} U(\Psi(f,p))dM(f)\right)d\mu(p) = 0$$

for all $L, M \in \Delta(\mathcal{H})$.

Suppose $\alpha < 0$. Then by taking $u = -1$ and $v(p) = 0$ for all $p \in \Delta(\Omega)$ we have

$$\alpha u - \int_{\Delta(\Omega)} v(p)d\mu(p) = -\alpha > 0,$$

which is a contradiction to the separation condition. Hence $\alpha \geq 0$.

Suppose $\alpha = 0$. Then

$$-\int_{\Delta(\Omega)}\left(\int_{\mathcal{H}} U(\Psi(f,p))dL(f) - \int_{\mathcal{H}} U(\Psi(f,p))dM(f)\right)d\mu(p) = 0$$

holds for all $L, M \in \Delta(\mathcal{H})$. This condition is violated when $L = \delta(l)$ and $M = \delta(m)$ with $U(l) > U(m)$. Hence $\alpha > 0$.

Now suppose $\mu = 0$. Then we have

$$\alpha\left(\int_{\mathcal{H}} U(f)dL(f) - \int_{\mathcal{H}} U(f)dM(f)\right) = 0$$

for all $L, M \in \Delta(\mathcal{H})$. But again this cannot be true for $L = \delta(l)$ and $M = \delta(m)$ with $U(l) > U(m)$. Hence μ is non-zero.

Suppose $\int_{\Delta(\Omega)} v(p)\mu(dp) < 0$ for some $v \in C_+(\Delta(\Omega))$. Then for sufficiently small $\varepsilon > 0$ we have

$$\alpha(-\varepsilon) - \int_{\Delta(\Omega)} v(p)d\mu(p) > 0,$$

which contradicts the separation condition. Hence $\mu \in (C_+(\Delta(\Omega)))^* = \mathcal{M}(\Delta(\Omega))$, where the last equality follows from Theorem 14.12 in Aliprantis and Border [3].

Hence without loss of generality we normalize $\alpha = 1$ and $\mu(\Delta(\Omega)) = 1$. Pick $M = \delta(m)$ with $U(m) = 0$ without loss of generality. Then by taking $L = \delta(f)$ we have

$$U(f) - \int_{\Delta(\Omega)} U(\Psi(f,p))d\mu(p) = 0.$$

Because U and u are ordinally equivalent over $\Delta(Z)$, there is a monotone function $\phi : u(\Delta(X)) \to \mathbb{R}$ such that

$$U(l) = \phi(u(l)).$$

Thus, by mixture-linearity of u, we have

$$\begin{aligned} U(f) &= \int_{\Delta(\Omega)} \phi\left(u\left(\Psi(f, p)\right)\right) d\mu(p) \\ &= \int_{\Delta(\Omega)} \phi\left(u\left(\bigoplus_{\omega\in\Omega} f(\omega)p(\omega)\right)\right) d\mu(p) \\ &= \int_{\Delta(\Omega)} \phi\left(\sum_{\omega\in\Omega} u\left(f(\omega)\right) p(\omega)\right) d\mu(p). \end{aligned}$$

□

Note that when ϕ is linear the function $U(f)$ reduces to the subjective expected utility form

$$U(f) = \sum_{\omega\in\Omega} u(f(\omega))p_\mu(\omega),$$

where

$$p_\mu = \int_{\Delta(\Omega)} p\mu(dp)$$

is the expected value of distribution over Ω according to μ.

6.6 The Model of Incomplete Ranking

Bewley [11] dropped the completeness axiom from the Savage/Anscombe-Aumann system and obtained the representation

$$f \succsim g \iff \sum_{\omega\in\Omega} u(f(\omega))p(\omega) \geq \sum_{\omega\in\Omega} u(g(\omega))p(\omega) \text{ for all } p \in P,$$

which I present in the finite-state setting, where P is a convex subset of $\Delta(\Omega)$. The interpretation is that the decision maker can rank act f over g only when all priors in P agree, which makes sense under ambiguity and ignorance.

Note, however, that this class of preferences satisfies the sure-thing principle/the independence axiom. Hence the Ellsberg paradox cannot be a direct motivation of this model. Gilboa et al. [47] showed that it can be related to the multiple-priors model (Gilboa and Schmeidler [45]) and the Ellsberg paradox (Ellsberg [28]) in the sense that the incomplete ranking model describes the decision maker's "objective rationality" and the multiple-priors model describes his/her "subjective rationality." The objective model describes the decision maker's unambiguous ranking, and the subjective one describes what determines his/her choice with ambiguity.

To illustrate, let $\succsim^*$ be the possibly incomplete ranking over the Anscombe-Aumann domain $\mathcal{H}$, which describes the decision maker's objective rationality and is represented in the Bewley form above. Also, let $\succsim^\wedge$ be the *complete* ranking over $\mathcal{H}$, which describes the decision maker's subjective rationality and falls in the multiple-priors model.

They impose two axioms which relate the two models:

1. **Consistency**: For all $f, g \in \mathcal{H}$, $f \succsim^* g$ implies $f \succsim^\wedge g$.
2. **Caution**: For all $f, g \in \mathcal{H}$, $g \not\succsim^* f$ implies $f \succsim^\wedge g$.

Note, the second axiom states that we have to be indifferent between two acts when we cannot rank between them.

Under the above two axioms, the two models are translatable to each other in the sense that the pair (u, P) forms a Bewley-type representation of $\succsim^*$ if and only if it forms a multiple-priors representation of $\succsim^\wedge$.

6.7 Information and Belief

The subjective expected utility theory and its generalization to allow ambiguity aversion state only that when preference over acts satisfies certain axioms there exists *some* subjective belief that forms a representation, whether it is a unique probability measure or a more general one, and stay silent about what such subjective belief should look like and how it should be related to information that is given prior to forming a belief.

Also, even when we allow ambiguity aversion, the theory cannot explain how ambiguity aversion arises. Is it because information is more ambiguous, or is it because the decision maker is inherently more pessimistic? The theory cannot tell. For, information is taken to be an implicit fixed factor in the subjective expected utility theory.

For example, in the maxmin expected utility model

$$U(f) = \min_{p \in P} \sum_{\omega \in \Omega} u(f(\omega))p(\omega)$$

we saw that the set P being larger (in the sense of set inclusion) corresponds to more ambiguity aversion. However, since P is a part of the representation of preference, we cannot tell if P being larger is due to more ambiguity of information or due to the decision maker being inherently more pessimistic.

To understand a functional relationship between information and subjective belief, we need to treat *objective imprecise probabilistic information* as an explicit variable. Note that if it is objective and precise we go back to the vNM expected utility theory.

Following Gajdos et al. [38], this section deals with information in the form of a set of probability distributions, such that you know the true distribution lies in it but do not know anything about which one it is true or which one in it is more likely to be true. We call such a set a *probability possibility set*, and treat it as a variable.

Let Ω be the set of states of the world, which is assumed to be finite. Let $\Delta(\Omega)$ denote the set of probability distributions over Ω, which is a compact convex subset of $\mathbb{R}^{\Omega}$. Let $\mathcal{P}$ denote the set of closed (hence compact) and convex subsets of $\Delta(\Omega)$, that is, the family of probability possibility set. When a probability possibility set $P \in \mathcal{P}$ is given, the decision maker knows that the true probability distribution is in P but does not know anything about which one in P is true or which one in it is more likely to be true. Note that when a probability possibility set is singleton set, say $\{p\}$, it reduces to the special case that the decision maker is given precise probabilistic information.

We assume that probability possibility sets are convex as well as compact. This is natural in the context where the set of possible probability distributions is characterized by linear inequalities.

Let Z be the set of pure outcomes and let $\Delta_S(Z)$ be the set of simple lotteries over Z.

An Anscombe-Aumann act (AA act) is a mapping from Ω into $\Delta_S(Z)$, or equivalently as an element of $\Delta_S(Z)^{\Omega}$. Let $\mathcal{H}$ be the set of AA acts.

The decision maker has preference over $\mathcal{P} \times \mathcal{H}$. That is, given $P, Q \in \mathcal{P}$ and $f, g \in \mathcal{H}$, the relation

$$(P, f) \succsim (Q, g)$$

states that choosing bet f under information P is at least as good as choosing bet g under information Q.

Another way to model preference will be to consider binary relation indexed by P in the form

$$f \succsim_P g,$$

and to consider a family $\{\succsim_P\}_{P\in\mathcal{P}}$, as in Hayashi [58].

This way requires less observation, but we dare to consider observing direct preference over information here. For, the following experiment is possible.

Example 6.2 Each of Box A and Box B contains 100 balls, which are either red or blue. The proportion of colors in Box A is totally unknown. There are 50 red balls and 50 blue balls in Box B.

Choose one from A and B. If red is drawn from the box you choose you get 100 dollars and nothing if blue is drawn from there.

It will be reasonable to choose B in the above. Then what if B contains 49 red balls? What if 48? And so on. You can continue asking such questions. At some point you will switch from B to A, since it will be reasonable to choose A over B if B contains norRed ball. Such a threshold, which we call probability equivalent, is going to be the measure of how the decision maker dislikes imprecise probabilistic information.

Keeping the above example in mind, let us consider axioms on preference over pairs of probability possibility sets and acts.

The first one will need no explanation in the current context.

Axiom 6.14 (Weak Order) $\succsim$ is complete and transitive.

Next, we assume that preference over acts follows the Gilboa-Schmeidler theory under any *fixed* information.

Axiom 6.15 (Mixture Continuity) For all $P \in \mathcal{P}$ and for all $f, g, h \in \mathcal{H}$, the sets

$$\{\alpha \in [0, 1] : (P, \alpha f + (1 - \alpha)g) \succsim (P, h)\}$$

and

$$\{\alpha \in [0, 1] : (P, h) \succsim (P, \alpha f + (1 - \alpha)g)\}$$

are closed subsets of $[0, 1]$.

Axiom 6.16 (Monotonicity) For all $P \in \mathcal{P}$ and for all $f, g \in \mathcal{H}$, if $(P, f(\omega)) \succsim (P, g(\omega))$ holds for all $\omega \in \Omega$, then $(P, f) \succsim (P, g)$.

Axiom 6.17 (Certainty Independence) For all $P \in \mathcal{P}$ and for all $f, g \in \mathcal{H}$ and $l \in \Delta_S(Z)$, $\lambda \in (0, 1]$,

$$(P, f) \succsim (P, g) \iff (P, \lambda f + (1 - \lambda)l) \succsim (P, \lambda g + (1 - \lambda)l).$$

Axiom 6.18 (Uncertainty Aversion) For all $P \in \mathcal{P}$ and for all $f, g \in \mathcal{H}$ and $\lambda \in [0, 1]$,

$$(P, f) \sim (P, g) \implies (P, \lambda f + (1 - \lambda)g) \succsim (P, f).$$

Also, it will be natural that imprecise probabilistic information affects choice only though affecting the decision maker's belief and does not affect taste over outcomes, in the current scope.

Axiom 6.19 (Outcome Preference) For all $P, Q \in \mathcal{P}$ and $l \in \Delta_S(Z)$,

$$(P, l) \sim (Q, l).$$

Also, for all $P \in \mathcal{P}$ there exist $l, m \in \Delta_S(Z)$ such that

$$(P, l) \succ (P, m).$$

Given the above conditions, we can provide a preliminary characterization.

Theorem 6.6 *The following two statements are equivalent:*

(a) $\succsim$ *satisfies Weak Order, Mixture Continuity, Monotonicity, Certainty Independence, Uncertainty Aversion and Outcome Preference.*
(b) There exists a mixture-linear function $u : \Delta_S(X) \to \mathbf{R}$ *and a mapping* $\varphi : \mathcal{P} \to \mathcal{P}$ *such that preference* $\succsim$ *is represented in the form*

$$U(P, h) = \min_{p \in \varphi(P)} \sum_{\omega \in \Omega} u(h(\omega))\ p(\omega).$$

Moreover, u is unique up to positive affine transformation and φ *is unique.*

Proof Fix any arbitrary $P \in \mathcal{P}$ and consider the binary relation

$$(P, f) \succsim (P, g),$$

then it satisfies the condition for the Gilboa-Schmeidler theorem. Hence there exist a set $\varphi(P) \in \mathcal{P}$ and a mixture-linear function $u_P : \Delta_S(X) \to \mathbf{R}$ such that the ranking conditional on P is represented in the form

$$\min_{p \in \varphi(P)} \sum_{\omega \in \Omega} u_P(h(\omega))\ p(\omega).$$

By Outcome Preference, without loss of generality we can fix u so that we have $u_P = u$ for all P. □

From the above result we obtain a mapping φ that maps each probability possibility set P into the corresponding set of priors $\varphi(P)$. Yet, there is no restriction on the relation between P and $\varphi(P)$. We impose two natural restrictions.

First is that as far as we are given precise information, the decision maker uses it to calculate the probability distribution over outcomes. Given $p \in \Delta(\Omega)$ and $f \in \mathcal{H}$, let $l(p, f) \in \Delta(X)$ be the lottery given by $l(p, f) = \sum_{\omega \in \Omega} f(\omega) p(\omega)$.

Axiom 6.20 (Reduction Under Precise Information) For all $p \in \Delta(\Omega)$ and $f \in \mathcal{H}$,

$$(\{p\}, f) \sim (\{p\}, l(p, f)).$$

Second is, given a probability possibility set P, if every possible probability distribution in it tells us that bet f is better than bet g, the decision maker would or should rank f over g.

Axiom 6.21 (Dominance) For all $P \in \mathcal{P}$ and for all $f, g \in \mathcal{H}$, if

$$(\{p\}, f) \succsim (\{p\}, g)$$

holds for all $p \in P$, then

$$(P, f) \succsim (P, g).$$

Note that Dominance does not imply Reduction under Precise Information. Here is an example: $\varphi(P) = \lambda P + (1 - \lambda)\{p^*\}$, where $p^* \in \Delta(\Omega)$ is a fixed probability distribution.

Under the two additional axioms, there is a natural restriction on φ.

Theorem 6.7 *The following two statements are equivalent:*

(a) $\succsim$ *satisfies Weak Order, Mixture Continuity, Monotonicity, Certainty Independence, Uncertainty Aversion, Outcome Preference, Reduction under Precise Information and Dominance.*
(b) There exists a mixture-linear function $u : \Delta_S(X) \to \mathbf{R}$ *and a mapping* $\varphi : \mathcal{P} \to \mathcal{P}$ *such that preference* $\succsim$ *is represented in the form*

$$U(P, h) = \min_{p \in \varphi(P)} \sum_{\omega \in \Omega} u(h(\omega))\, p(\omega),$$

where φ *satisfies*

$$\varphi(P) \subset P$$

for all $P \in \mathcal{P}$.

Moreover, u *is unique up to positive affine transformation and* φ *is unique.*

Proof From Reduction under Precise Information we have $\varphi(\{p\}) = \{p\}$.

Now suppose there is some $P \in \mathcal{P}$ such that $\varphi(P) \not\subset P$, and pick $p \in \varphi(P) \setminus P$. Then by the separating hyperplane theorem there exist a vector $x \in \mathbb{R}^S \setminus \{\mathbf{0}\}$ and a number α such that

$$\langle x, q\rangle \geq \alpha > \langle x, p\rangle, \quad \forall q \in P$$

holds.

Without loss of generality, say that we can take $l \in \Delta(X)$ with $u(l) = \alpha$, and $f \in \mathcal{H}$ with $u \circ f = x$. Then the above condition leads to

$$\langle u \circ f, q \rangle \geq u(l) > \langle u \circ f, p \rangle, \quad \forall q \in P.$$

From the left half of the condition,

$$(\{q\}, f) \succsim (\{q\}, l)$$

holds for all $q \in P$. Hence by Dominance, $(P, f) \succsim (P, l)$. On the other hand, however, from the right half of the condition we obtain

$$u(l) > \langle u \circ f, p \rangle \geq \min_{p' \in \varphi(P)} \langle u \circ f, p' \rangle,$$

meaning $(P, l) \succ (P, f)$, a contradiction. □

Note that there is still no restriction on how the decision maker will respond to *changes* in probability possibility sets. This motivates us to consider the final three axioms.

First is a mixture-independence axiom over the space of probability possibility sets, in the spirit of von-Neumann and Morgenstern. Given $P, Q \in \mathcal{P}$ and $\lambda \in [0, 1]$, define the mixture by

$$\lambda P + (1 - \lambda) Q = \{\lambda p + (1 - \lambda) q : p \in P, q \in Q\}$$

Interpret this mixture set as information saying "the true probability distribution lies in P with probability λ and lies in Q with probability $1 - \lambda$," then we can motivate the independence axiom based on the idea of dynamic consistency as before.

Note, there are two "cheats" here. One is identifying $\lambda p + (1 - \lambda) q$ with a probabilistic process delivering "probability distribution p with probability λ and probability distribution q with probability $1 - \lambda$," which is familiar. The other is identifying information saying "the true probability distribution lies in P with probability λ and lies in Q with probability $1 - \lambda$" with information saying "the true probability distribution lies in $\lambda P + (1 - \lambda) Q$."

Having said so, the mixture-independence axiom takes the following form.

Axiom 6.22 (Information Independence) For all $f \in \mathcal{H}$, for all $P, Q, R \in \mathcal{P}$ and $\lambda \in [0, 1]$,

$$(P, f) \succsim (Q, f) \implies (\lambda P + (1 - \lambda) R, f) \succsim (\lambda Q + (1 - \lambda) R, f).$$

Second is a symmetry axiom, based on the idea of principle of insufficient reasons. For this purpose, we begin with defining "symmetric" transformation of probability distributions.

Definition 6.3 An $S \times S$ matrix Π is doubly stochastic if all its elements are non-negative and

$$\sum_{k=1}^{S} \Pi_{kl} = 1, \; \sum_{l=1}^{S} \Pi_{kl} = 1$$

holds for all $k, l \in S$.

A doubly stochastic matrix makes a linear transformation that maps a probability distribution into a probability distribution.

Definition 6.4 A doubly stochastic matrix Π is said to be a unitary transformation if

$$(\{p\}, f) \succsim (\{q\}, f) \implies (\{\Pi p\}, f) \succsim (\{\Pi q\}, f)$$

holds for all $p, q \in \Delta(S)$ and $f \in \mathcal{H}$.

A unitary transformation keeps the rankings over probability distributions unchanged, and hence is seen as a natural "symmetric transformation."

The next axiom states that such transformation keeps the rankings over imprecise probability possibility sets, as it keeps the rankings over precise probability distributions. For arbitrary $P \in \mathcal{P}$ and unitary transformation Π, let $\Pi P = \{\Pi p : p \in P\}$.

Axiom 6.23 (Invariance to Unitary Transformations) For all unitary transformation Π,

$$(\Pi P, f) \succsim (\Pi Q, f) \implies (\Pi P, f) \succsim (\Pi Q, f)$$

holds for for all $P, Q \in \mathcal{P}$ and $f \in \mathcal{H}$.

Finally, we impose the condition that preference over probability possibility sets is continuous in the Hausdorff metric.

Axiom 6.24 (Information Continuity) For any sequences of probability possibility sets $\{P^\nu\}$ and $\{Q^\nu\}$ which converge to P and Q respectively in the Hausdorff metric, for all $f \in \mathcal{H}$,

$$(P^\nu, f) \succsim (Q^\nu, f), \; \forall \nu \implies (P, f) \succsim (Q, f).$$

This is a technical axiom but excludes an important example of belief selection, the center of gravity. The center of gravity of a triangle divides the midlines by 2 versus 1 and it is unchanged as far as the triangle remains to be a triangle. But consider that the triangle converges to a segment (imagine that B is getting closer to C in triangle ABC), then the center of gravity does not converges to the midpoint of the segment, which is the natural "center of gravity" there.

The current representation theorem leads us to select so-called Steiner point as the center.

Definition 6.5 Let $V = \{v \in \mathbb{R}^{\Omega} : \|v\| = 1,\ p \cdot \mathbf{1} = 0\}$ denote the intersection of the unit surface in the $|\Omega|$-dimensional Euclidian space and the plane with normal vector **1**.

Then the Steiner point of a probability possibility set $P \in \mathcal{P}$ is defined by

$$s(P) = \int_V \arg\max_{p \in P} p \cdot v \lambda(dv),$$

where λ is the uniform distribution probability measure over V.

For example, the Steiner point of a triangle is the weighted average of vertices in which the weight on each vertex is proportional to its outer angle.

Theorem 6.8 *The following two statements are equivalent:*

(a) $\succsim$ *satisfies Weak Order, Mixture Continuity, Monotonicity, Certainty Independence, Uncertainty Aversion, Outcome Preference, Reduction under Precise Information, Dominance, Information Independence, Invariance to Unitary Transformations and Information Continuity.*

(b) *There exist a mixture-linear function* $u : \Delta_S(X) \to \mathbb{R}$ *and a number* $\varepsilon \in [0, 1]$ *such that preference* $\succsim$ *is represented in the form*

$$U(P, h) = \min_{p \in \varphi(P)} \sum_{\omega \in \Omega} u(h(\omega))\ p(\omega),$$

where φ *has the form*

$$\varphi(P) = (1 - \varepsilon)\{s(P)\} + \varepsilon P$$

for all $P \in \mathcal{P}$.

Moreover, u is unique up to positive affine transformation and ε *is unique.*

Proof I give just a sketch of the proof here. See GHTV [38] for details.

By Information Independence, we obtain a mixture-linearity condition on φ,

$$\varphi(\lambda P + (1 - \lambda)Q) = \lambda\varphi(P) + (1 - \lambda)\varphi(Q).$$

By Invariance to Unitary Transformations we obtain

$$\varphi(\Pi P) = \Pi\varphi(P)$$

for all unitary transformation Π

Information Continuity implies that φ is continuous in the Hausdorff metric.

One can show that a mapping φ satisfying the property $\varphi(P) \subset P$ as well as the above three properties must have the form

$$\varphi(P) = (1 - \varepsilon)\{s(P)\} + \varepsilon P,$$

by applying Schneider's theorem [93]. □

Here we call the parameter ε the decision maker's degree of *imprecision aversion*. Given an objective probability possibility set P, the decision maker shrinks the set toward its Steiner point $s(P)$ according to the degree of imprecision aversion. For example, when $\varepsilon = 0$ the decision maker gives $\varphi(P) = \{s(P)\}$, where he always selects a single point, the Steiner point, as his belief. On the other hand, when $\varepsilon = 1$ the decision maker does not contract at all and gives $\varphi(P) = P$.

The degree of imprecision aversion can be measured by experiment. Coming back to Example 6.2, the Steiner point of probability possibility set for Box A is $(R, B) = (0.5, 0.5)$. Denote the probability equivalent of this set by p, then we obtain

$$\varepsilon = \frac{0.5 - p}{0.5}.$$

Chapter 7
Updating Belief and Dynamic Consistency

Suppose you have acquired information in the form of an event, meaning "I know that the true state of the world lies in this subset but do not know which one in it is true." Then do you update your subjective belief? In probability theory we normally follow the Bayes rule, but how does it have to be? Here we will see that Bayesian updating is a *necessary* requirement following from Dynamic Consistency.

7.1 Updating Subjective Belief

Here we follow Ghirardato [41], while the idea dates back at least to Epstein and Le Breton [30]. The set of choice alternatives is given by $\mathcal{F} = \{f : \Omega \to X\}$. Note that a kind of consequentialism is presumed here, but I will discuss this point later.

Let us consider the following two-period model. Let $\succsim$ denote the ex-ante preference over $\mathcal{F}$, and let $\{\succsim_E\}$ denote the family of ex-post preferences. For each event $E \subset \Omega$, the ranking $\succsim_E$ describes the decision maker's preference over $\mathcal{F}$ provided that he knows the true state belongs to E.

Given $E \subset \Omega$ and $f, g \in \mathcal{F}$, define $fEg \in \mathcal{F}$ by

$$fEg(\omega) = \begin{cases} f(\omega), & \text{if } \omega \in E \\ g(\omega), & \text{if } \omega \notin E \end{cases}.$$

We impose the following axioms.

Axiom 7.1 (Weak Order) $\succsim$ and $\succsim_E$ for all $E \subset \Omega$ satisfy completeness and transitivity.

Axiom 7.2 (Dynamic Consistency) For all $E \subset \Omega$ and $f, g \in \mathcal{F}$,

(i) $f \succsim_E g \implies fEg \succsim g$;
(ii) $fEg \succsim g \implies f \succsim_E g$.

T. Hayashi, *Decision Theory*, Monographs in Mathematical Economics 9,
https://doi.org/10.1007/978-981-95-2200-2_7

(i) states that if the decision maker prefers f over g ex-post after knowing E he prefers replacing g by f conditional on E ex-ante. Otherwise, an inconsistency happens in the sense that even if he did not choose to replace g by f conditional on E ex-ante he changes his mind and switches to prefer f ex-post once E occurs. Similarly for (ii).

Axiom 7.3 (Outcome Preference) For all $E \subset \Omega$ and $x, y \in X$,

$$x \succsim y \iff x \succsim_E y.$$

This means that an event only has the role of information and does not affect ranking over outcomes.

Axiom 7.4 (Weak Comparative Probability) For all $A, B \subset \Omega$ and $x^*, x, y^*, y \in X$ with $x^* \succ x$ and $y^* \succ y$,

$$\begin{bmatrix} x^* \text{ if } \omega \in A \\ x \text{ if } \omega \notin A \end{bmatrix} \succsim \begin{bmatrix} x^* \text{ if } \omega \in B \\ x \text{ if } \omega \notin B \end{bmatrix}$$

$$\Longrightarrow \begin{bmatrix} y^* \text{ if } \omega \in A \\ y \text{ if } \omega \notin A \end{bmatrix} \succsim \begin{bmatrix} y^* \text{ if } \omega \in B \\ y \text{ if } \omega \notin B \end{bmatrix}.$$

This axiom plays the same role as in the Savage theory. Similarly for the next two axioms.

Axiom 7.5 (Nontriviality) There exist $x, y \in X$ such that $x \succ y$.

Axiom 7.6 (Continuity) For all $f \succ g$ and $x \in X$, there is $\{E_k\}_{k=1}^n$, a partition of Ω, such that for all k,

$$x E_k f \succsim g$$

and

$$f \succ x E_k g.$$

The last axiom describes the standpoint that "once event E is realized what would have been obtained if E were not realized does not matter in the ex-post decision making."

Axiom 7.7 (Conditional Preference) For all $E \subset \Omega$ and $f, g \in \mathcal{F}$, if $f(\omega) = g(\omega)$ for all $\omega \in E$ then $f \sim_E g$.

The lemma below follows immediately.

Lemma 7.1 *Suppose that* $\succsim$ *and* $\{\succsim_E\}$ *satisfy Weak Order and Conditional Preference. Then Dynamic Consistency is met if and only if for all* $E \subset \Omega$ *and*

$f, g, h \in \mathcal{F}$,

$$fEh \succsim gEh \iff fEh \succsim_E gEh.$$

Proof Suppose $fEh \succsim gEh$. From Dynamic Consistency (ii) we have $f \succsim_E gEh$. Since $f \sim_E fEh$ follows from Conditional Preference, we obtain $fEh \succsim_E gEh$. To show the other direction, suppose $fEh \succsim_E gEh$. Then from Dynamic Consistency (i) we obtain $fEh \succsim gEh$. □

We will show that Eventwise Separability, Savage's P2, is rather a consequence of dynamic consistency under the mild axioms.

Axiom 7.8 (Eventwise Separability) For all $E \subset \Omega$ and $f, g, h, h' \in \mathcal{F}$,

$$fEh \succsim gEh \iff fEh' \succsim gEh'.$$

Proposition 7.1 *Under Weak Order and Conditional Preference, Dynamic Consistency implies Eventwise Separability.*

Proof Assume Dynamic Consistency. Then, for all $E \subset \Omega$ and $f, g, h, h' \in \mathcal{F}$, it follows from the previous lemma that $fEh \succsim gEh$ is equivalent to $fEh \succsim_E gEh$. From Conditional Preference this is equivalent to $fEh' \succsim_E gEh'$, and again it is equivalent to $fEh' \succsim gEh'$ from the previous lemma. □

Thus we have:

Theorem 7.1 *The two statements below are equivalent:*

(i) $\succsim$, $\{\succsim_E\}$ *satisfy Axioms 7.1–7.7.*
(ii) There exist a function u, a probability measure p and a family of probability measures $\{p_E\}$ such that $\succsim$ is represented in the form

$$\int_\Omega u(f(\omega))p(d\omega)$$

and for each non-null E the ranking $\succsim_E$ is represented in the form

$$\int_E u(f(\omega))p_E(d\omega)$$

and p_E follows the Bayes rule:

$$p_E(A) = \frac{p(E \cap A)}{p(E)}.$$

Proof It is clear that (i) is necessary for (ii), so we show sufficiency. It follows from Axioms 7.1–7.7 and Proposition 7.1 that $\succsim$, $\{\succsim_E\}$ have subjective expected utility representations. It remains to show that the Bayes rule must be met.

Since

$$
\begin{aligned}
f \succsim_E g &\Longleftrightarrow fEh \succsim gEh \\
&\Longleftrightarrow \int_E u \circ f dp + \int_{\Omega \setminus E} u \circ h dp \geqq \int_E u \circ g dp + \int_{\Omega \setminus E} u \circ h dp \\
&\Longleftrightarrow \frac{1}{p(E)} \int_E u \circ f dp \geqq \frac{1}{p(E)} \int_E u \circ g dp,
\end{aligned}
$$

we see that $\frac{1}{p(E)} \int_E u \circ f dp$ gives a subjective expected utility representation of $\succsim_E$. Hence we obtain the result from uniqueness of representation. □

Here are two kids of "consequentialism" assumed. One is already imposed by the setting, where the decision maker is indifferent in timing of resolution of uncertainty and cares only for (subjective) distribution of final outcomes. The other is imposed by Conditional Preference, which states that the interim decision after knowing an event is independent of what would have been obtained if such event did not happen.

Note also that the dynamic consistency condition is imposed on *any* event in the above version. One may think of weakening it by imposition the condition only on a *fixed* information structure. See the next section.

7.2 Updating Ambiguous Beliefs

Here we follow Epstein and Schneider [31] with simplification of the setting so that there is no intermediate consumption.

ES considered a fixed information structure, since we already know that imposing dynamic consistency over any event implies event separability. Perhaps fixed information structure is also the reality, when we can vary information structure only in a hypothetical manner. Note that consequentialism is still maintained.

We consider a two-period model to illustrate. Let Ω be the set of states, which is assumed to be finite. Information structure is described as a partition of the state space $\{E_k\}_{k=1}^n$ at Period 0 the decision maker knows nothing and knows some event E_k at Period 1 and uncertainty resolves completely at Period 2.

We consider the domain of Anscombe-Aumann acts $\mathcal{H} = \{h : \Omega \to \Delta(X)\}$. Denote the ex-ante preference held at Period 0 by $\succsim$, and the sequence of preferences at Period 1 by $\{\succsim_{E_k}\}_{k=1}^n$, where $\succsim_{E_k}$ refers to the preference over $\mathcal{H}$ that is held after knowing E_k.

Now one might think that we can simply extend the Gilboa-Schmeidler's representation in a straightforward manner. That is, we might consider that the ex-ante preference $\succsim$ is represented in the form

$$
\min_{p \in P} \sum_{\omega \in \Omega} u(f(\omega)) p(d\omega)
$$

and the interim preference $\succsim_E$ after knowing E is represented in the form

$$\min_{p_E \in P_E} \sum_{\omega \in E} u(f(\omega)) p_E(d\omega),$$

where the set of posteriors P_E is obtained by **prior-by-prior Bayesian updating**:

$$P_E = \left\{ p_E : p_E(A) = \frac{p(A \cap E)}{p(E)},\ p \in P \right\}.$$

Without any restriction, the above leads to dynamic inconsistency in general. Let us look at the following example.

Example 7.1 Consider that the state space is $\Omega = \{\omega_1, \omega_2, \omega_3\}$ and the set of priors is given by $P = \{p : p(\omega_1) = \frac{1}{3}, \frac{1}{12} \leq p(\omega_2) \leq \frac{1}{6}\}$. In the next period the decision maker knows either $\{\omega_2\}$ or $\{\omega_1, \omega_3\}$.

If he knows $\{\omega_2\}$ the set of posteriors is simply $P_{\{\omega_2\}} = \{p : p(\omega_2) = 1\}$. If he knows $\{\omega_1, \omega_3\}$ the set of posteriors is $P_{\{\omega_1, \omega_3\}} = \{p : p(\omega_2) = 0,\ \frac{4}{11} \leq p(\omega_1) \leq \frac{2}{5}\}$. The funny thing is that he knew $p(\omega_1) = \frac{1}{3}$ precisely but now he knows $p(\omega_1)$ only in an imprecise way.

Consider the following two bets:

f: 160 if ω_1, nothing otherwise.
g: 100 if ω_3, nothing otherwise.

Assume risk-neutrality for simplicity, or that payoffs are already adjusted to risk attitudes.

Then, at Period 0 we have $f \succ g$ since $160 \cdot \frac{1}{3} > 100 \cdot \frac{6}{12}$. However, once $\{\omega_1, \omega_3\}$ is known to the decision maker we have $g \succ f$ since $160 \cdot \frac{4}{11} < 100 \cdot \frac{6}{10}$.

Note that f and g are identical on ω_2, and non-occurrence of ω_2 does not change anything about how f and g differ in terms of outcomes after knowing $\{\omega_1, \omega_3\}$.

Example 7.2 There are two coins. They may be perfectly correlated or perfectly negatively correlated, and you don't know anything about which is true or which is more likely to be true, and also you cannot exclude anything between the two.

The state space is described as $\Omega = \{H_1H_2, H_1T_2, T_1H_2, T_1T_2\}$, where H_1H_2 refers to heads in the first coin and heads in the second coin, and similarly for the others.

Note, however, that the ex-ante probability of each coin flip is even-chances. That is, the set of priors is $P = \{p : p(H_1) = p(T_1) = 0.5,\ p(H_2) = p(T_2) = 0.5\}$.

Now suppose that the decision maker is told that the first flip was heads. Then the set of posteriors obtained by prior-by-prior Bayesian updating is $P_{H_1} = \{p : 0 \leq p(H_2) \leq 1\}$, because, if they are perfectly correlated it must be $p(H_2) = 1$, and if they are perfectly negatively correlated it must be $p(H_2) = 0$, and anything between them is possible. That is, despite that you knew the ex-ante probability of the second flip is even chances now you know nothing once you know the first flip. Similarly when the first flip was tails.

Now consider the following two bets:

f: 100 if the second flip is heads, nothing otherwise.
g: $50 - \varepsilon$ for sure.

Again assume risk-neutrality for simplicity. Then, despite that we have $f \succ g$ ex-ante, once the first flip is known we have $g \succ_{H_1} f$ and $g \succ_{T_1} f$ in each possible case.

The above example tells us that for dynamic consistency to be met the system of sets of beliefs must have a certain structure.

To think of this problem, let us extend the Gilboa-Schmeidler axioms in the following way. Again, let me emphasize that the information structure is fixed.

Axiom 7.9 (Conditional Preference) For all k and $f, g \in \mathcal{H}$, if $f(\omega) = g(\omega)$ for all $\omega \in E_k$ then $f \sim_{E_k} g$.

Axiom 7.10 (GS) $\succsim$ and $\{\succsim_{E_k}\}_{k=1}^n$ satisfy the Gilboa-Scchmeidler axioms.

Axiom 7.11 (Dynamic Consistency) For all $f, g \in \mathcal{H}$, if $f \succsim_{E_k} g$ for all $k = 1, cdots, n$ then $f \succsim g$.

The extended Gilboa-Schmeidler axioms together with Conditional Preference and Dynamic Consistency characterize the following structure.

Definition 7.1 Given an information structure $\{E_k\}_{k=1}^n$, the sequence of sets of beliefs $P, \{P_{E_k}\}_{k=1}^n$ is said to satisfy rectangularity if

$$P = \left\{ \sum_{k=1}^{n} p(E_k) p_{E_k} : p \in P,\ p_{E_k} \in P_{E_k},\ k = 1, \cdots, n \right\}.$$

Theorem 7.2 (Epstein-Schneider) *The two statements below are equivalent:*

(i) $\succsim, \{\succsim_{E_k}\}_{k=1}^n$ *satisfies Conditional Preference, Dynamic Consistency and GP.*
(ii) The ex-ante preference $\succsim$ *is represented in the form*

$$U(f) = \min_{p \in P} \sum_{\omega \in \Omega} u(f(\omega)) p(\omega),$$

and for each k the conditional preference $\succsim_{E_k}$ *is represented in the form*

$$U_{E_k}(f) = \min_{p_{E_k} \in P_{E_k}} \sum_{\omega \in E_k} u(f(\omega)) p_{E_k}(\omega),$$

and $(P, \{P_{E_k}\}_{k=1}^n)$ *satisfy rectangularity.*

Moreover, U and $\{U_k\}_{k=1}^n$ satisfy the recursive formula

$$U(f) = \min_{p \in P} \sum_{k=1}^{n} U_{E_k}(f) p(E_k).$$

Proof Since (i) is clearly necessary for (ii), we show sufficiency.

Since we already know that the GS axiom allows us to have maxmin representations for $\succsim$, $\{\succsim_{E_k}\}_{k=1}^n$, it suffices to show rectangularity and the recursive formula.

Pick any $f \in \mathcal{H}$, and for each $k = 1, \cdots, n$ pick l_k such that $f \sim_{E_k} l_k$. Let $\widetilde{f}$ denote the act given by such evenly-contingent outcomes. Then, by construction

$$U_{E_k}(f) = \min_{p_{E_k} \in P_{E_k}} \sum_{\omega \in E_k} u(f(\omega)) p_{E_k}(\omega) = U_{E_k}(\widetilde{f})$$

holds for each $k = 1, \cdots, n$.

Since $f \sim \widetilde{f}$ follows from Dynamic Consistency, we have $U(f) = U(\widetilde{f})$.

On the other hand, since all uncertainty is resolved at Period 1 regarding act $\widetilde{f}$,

$$U(\widetilde{f}) = \min_{p \in P} \sum_{k=1}^{n} u(l_k) p(E_k) = \min_{p \in P} \sum_{k=1}^{n} U_{E_k}(\widetilde{f}) p(E_k).$$

Therefore, for general f we obtain the recursive formula

$$U(f) = \min_{p \in P} \sum_{k=1}^{n} U_{E_k}(f) p(E_k).$$

Since this is rewritten into

$$\begin{aligned}
U(f) &= \min_{p \in P} \sum_{k=1}^{n} \left\{ \min_{p_{E_k} \in P_{E_k}} \sum_{\omega \in E_k} u(f(\omega)) p_{E_k}(\omega) \right\} p(E_k) \\
&= \min_{p \in P} \min_{p_{E_1} \in P_{E_1}} \cdots \min_{p_{E_n} \in P_{E_n}} \sum_{k=1}^{n} \left\{ \sum_{\omega \in E_k} u(f(\omega)) p_{E_k}(\omega) \right\} p(E_k) \\
&= \min_{q \in \left\{ \sum_{k=1}^{n} p(E_k) p_{E_k} : p \in P,\ p_{E_k} \in P_{E_k},\ k=1,\cdots,n \right\}} \sum_{\omega \in \Omega} u(f(\omega)) q(\omega)
\end{aligned}$$

we obtain rectangularity because of uniqueness of representation. □

Chapter 8
Intertemporal Choice

8.1 Streams Under Certainty

We follow the classic arguments by Koopmans [68, 70].

Consider an infinite and discrete time horizon $T = \{1, 2, \cdots\}$. Let C denote the set of per-period outcomes such as consumptions, which is assumed to be a compact and metric space. Then the set of all possible infinite streams of outcomes is given by C^∞. Note that C^∞ is compact metric with regard to the product metric.

In intertemporal situations, preference at a given period depends on past histories. Habit formation and addiction are such examples. Thus, let $\mathbf{c}^{t-1} = (c_0, \cdots, c_{t-1})$ denote a history of outcomes before Period t. Note that history before Period 1, $\mathbf{c}^0$, is null.

When $\mathbf{c}^{t-1}$, a history before Period t is given, denote preference defined over infinite streams starting at Period t by $\succsim_{\mathbf{c}^{t-1}}$. That is, when the relation holds $\mathbf{c}_t \succsim_{\mathbf{c}^{t-1}} \mathbf{c}'_t$ between $\mathbf{c}_t = (c_t, c_{t+1}, \cdots) \in C^\infty$ and $\mathbf{c}'_t = (c'_t, c'_{t+1}, \cdots) \in C^\infty$ it says that the individual after history $\mathbf{c}^{t-1}$ prefers "receiving c_t today, c_{t+1}, and so on" over "receiving c'_t today, c'_{t+1} tomorrow, and so on."

We consider that the decision maker has such preference at each period and after each possible history. Such a *process* is now denoted by $\{\succsim_{\mathbf{c}^{t-1}}\}$.

The following Dynamic Consistency is the most fundamental axiom in this context.

Axiom 8.1 (Dynamic Consistency) For all $t \in T$ and for all $\mathbf{c}^{t-1}$,

$$(c_t, \mathbf{c}_{t+1}) \succsim_{\mathbf{c}^{t-1}} (c_t, \mathbf{c}'_{t+1}) \iff \mathbf{c}_t \succsim_{(\mathbf{c}^{t-1}, c_t)} \mathbf{c}'_{t+1}$$

holds for all $c_t \in C$ and $\mathbf{c}_{t+1}, \mathbf{c}'_{t+1} \in C^\infty$, where $(\mathbf{c}^{t-1}, c_t) = (c_1, \cdots, c_{t-1}, c_t)$ denotes the augmented history after having c_t following $\mathbf{c}^{t-1}$.

T. Hayashi, *Decision Theory*, Monographs in Mathematical Economics 9,
https://doi.org/10.1007/978-981-95-2200-2_8

Dynamic Consistency states that there is no contradiction between decisions across histories. To understand, suppose for example that we have $\mathbf{c}_{t+1} \prec_{(\mathbf{c}^{t-1}, c_t)} \mathbf{c}'_{t+1}$ despite $(c_t, \mathbf{c}_{t+1}) \succsim_{\mathbf{c}^{t-1}} (c_t, \mathbf{c}'_{t+1})$.

Then, despite that this decision maker prefers $\mathbf{c}_{t+1}$ at least weakly over $\mathbf{c}'_{t+1}$ at Period t under history $\mathbf{c}^{t-1}$ *as a plan starting tomorrow* following the given current outcome c_t, once tomorrow comes and he faces an actual history $(\mathbf{c}^{t-1}, c_t)$ he changes his mind and prefers $\mathbf{c}'_{t+1}$ over $\mathbf{c}_{t+1}$. When this happens, the decision maker's "future self" will not follow the plan made by the decision maker's "current self." This makes it impossible to describe a decision maker's life course as a solution to a *planning problem*.

How should we describe the decision maker's choice when there is dynamic inconsistency? We will come to this issue in a later section.

The next axiom states that preference at each period is independent of past histories, which rule out habit formation and addiction. It is empirically a problem of course, but it makes a point that a human being does not change.

Axiom 8.2 (History Independence) For all $t \in T$ and $\mathbf{c}^{t-1}, \tilde{\mathbf{c}}^{t-1}$

$$\mathbf{c}_t \succsim_{\mathbf{c}^{t-1}} \mathbf{c}'_t \iff \mathbf{c}_t \succsim_{\tilde{\mathbf{c}}^{t-1}} \mathbf{c}'_t$$

holds for all $\mathbf{c}_t, \mathbf{c}'_t \in C^\infty$

Since preference is independent of histories under the above axiom we would simply write $\succsim_t$ as preference at Period t instead of $\succsim_{\mathbf{c}^{t-1}}$. Thus we denote a sequence of (possibly time-dependent) preferences by $\{\succsim_t\}_{t=1}^\infty$.

The next axiom states that a sequence of preferences $\{\succsim_t\}_{t=1}^\infty$ is invariant to time, that is, preferences are identical at every period.

Axiom 8.3 (Time Invariance) For all $t, s \in T$,

$$\mathbf{c} \succsim_t \mathbf{c}' \iff \mathbf{c} \succsim_s \mathbf{c}'$$

holds for all $\mathbf{c}, \mathbf{c}\prime \in C^\infty$.

Note that in the relation $\mathbf{c} \succsim_t \mathbf{c}'$ the sequences $\mathbf{c}$ and $\mathbf{c}'$ are seen as starting at Period t, whereas they are seen as starting at Period s in the relation $\mathbf{c} \succsim_s \mathbf{c}'$, where we shift time indices suitably. We adopt this convention for simplicity of notation.

Time Invariance, in addition to History Independence, is of course a strong restriction but it is also realistic. Think of the following example of a sequence of preferences:

Day 1: Today is special, so I eat a lot. From tomorrow, I will be moderate.
Day 2 and onward: The special day is over, so I will be moderate from today and forever.

This sequence/process *is* dynamically consistent, but do you think it is realistic? In a sense, Dynamic Consistency is a vacuous condition which can be met simply by following the "projection" of the initial preference. It will rather be empirically plausible, however, that once you say "today is special" you will say the same thing again tomorrow and onward. Or, if you exhibit some kind of impatience today it would be natural to assume that you exhibit the same kind of impatience again tomorrow and onward. Perhaps it is a cynical view, but this is what Time Invariance says.

Now under Time Invariance we reduce the argument to a single preference denoted by $\succsim$, instead of a sequence $\{\succsim_t\}_{t=0}^{\infty}$.

Given a single preference $\succsim$, we consider the following axiom called Stationarity.

Axiom 8.4 (Stationarity) For all $c_1 \in C$ and $\mathbf{c}, \mathbf{c}' \in C^{\infty}$,

$$(c_1, \mathbf{c}) \succsim (c_1, \mathbf{c}') \iff \mathbf{c} \succsim \mathbf{c}'$$

Provided that today's consequence is the same, compare two streams starting tomorrow. Stationarity says that if a sequence is better as a plan starting tomorrow it is better as a plan starting today as well. While it is no more than a property of a single preference, it is conventional to take it as the synonym of Dynamic Consistency when History Independence and Time Invariance are assumed.

Proposition 8.1 *Under History Independence and Time Invariance, Dynamic Consistency and Stationarity are equivalent.*

Proof Because of History Independence the preference process is written in the form $\{\succsim_t\}_{t=1}^{\infty}$.

Now, for two arbitrary consecutive preferences $\succsim_t, \succsim_{t+1}$, when the Dynamic Consistency condition

$$(c, \mathbf{c}) \succsim_t (c, \mathbf{c}') \iff \mathbf{c} \succsim_{t+1} \mathbf{c}'$$

is met, since the right-hand side is replaced by $\mathbf{c} \succsim_t \mathbf{c}'$ because of Time Invariance, it is equivalent to the stationarity condition

$$(c, \mathbf{c}) \succsim_t (c, \mathbf{c}') \iff \mathbf{c} \succsim_t \mathbf{c}'.$$

Because of Time Invariance this holds for all t. □

Thus, hereafter we assume a single preference ranking $\succsim$ defined over C^{∞}, satisfying Stationarity, where streams are assumed to start at Period 1 without loss of generality.

We introduce two separability axioms.

Axiom 8.5 (Separability I) For all $c_1, c_1' \in C$ and $\mathbf{c}, \mathbf{c}' \in C^{\infty}$,

(a)

$$(c_1, \mathbf{c}) \succsim (c_1, \mathbf{c}') \iff (c_1', \mathbf{c}) \succsim (c_1', \mathbf{c}')$$

(b)

$$(c_1, \mathbf{c}) \succsim (c_1', \mathbf{c}) \iff (c_1, \mathbf{c}') \succsim (c_1', \mathbf{c}').$$

Condition (a) states that preference over streams starting tomorrow does not depend on outcome today. Condition (b) states that preference over outcomes today does not depend on streams starting tomorrow.

The version of Stationarity we introduced above actually implies Separability (a), but we listed it above as it makes sense by itself.

Lemma 8.1 *Stationarity implies Separability (a).*

Proof The condition $(c_1, \mathbf{c}) \succsim (c_1, \mathbf{c}')$ is equivalent to $\mathbf{c} \succsim \mathbf{c}'$ because of Stationarity. By Stationarity again it is equivalent to $(c_1', \mathbf{c}) \succsim (c_1', \mathbf{c}')$. □

Theorem 8.1 *Fix a continuous representation U of preference $\succsim$ over C^∞ which satisfies completeness, transitivity and continuity.*

Then, $\succsim$ satisfies Stationarity if and only if there is a continuous function $W : C \times U(C^\infty) \to \mathbb{R}$ being strongly increasing in the second argument such that U has the form

$$U(c_1, \mathbf{c}) = W(c_1, U(\mathbf{c})).$$

Moreover, if another pair $(\widehat{U}, \widehat{W})$ represents the same ranking as (U, W) in the above form, there is a monotone function $W : U(C^\infty) \to \mathbb{R}$ such that

$$\widehat{U}(\mathbf{c}) = f(U(\mathbf{c}))$$

for all $\mathbf{c} \in C^\infty$ and

$$\widehat{W}(c_1, f(V)) = f(W(c_1, V))$$

for all $c_1 \in C$ and $V \in U(C^\infty)$.

Proof We show sufficiency of Stationarity.

Since U is a representation of $\succsim$,

$$U(\mathbf{c}) \geqq U(\mathbf{c}') \iff \mathbf{c} \succsim \mathbf{c}'$$

holds for any $\mathbf{c}, \mathbf{c}' \in C^\infty$.

On the other hand, from Stationarity,

$$U(c_1, \mathbf{c}) \geqq U(c_1, \mathbf{c}') \iff (c_1, \mathbf{c}) \succsim (c_1, \mathbf{c}')$$

holds for any $c_1 \in C$, hence $U(\cdot)$ and $U(c_1, \cdot)$ represent the same preference $\succsim$ over C^∞. Therefore there exists a monotone transformation depending on c_1, namely

$W(c_1, \cdot) : U(C^\infty) \to \mathbb{R}$, such that

$$U(c_1, \cdot) = W(c_1, U(\cdot)).$$

To show the second statement, suppose another pair $(\widehat{U}, \widehat{W})$ represents the same ranking as (U, W) in the above form. Then, by ordinal equivalence there is a monotone function $W : U(C^\infty) \to \mathbb{R}$ such that

$$\widehat{U}(\mathbf{c}) = f(U(\mathbf{c}))$$

for all $\mathbf{c} \in C^\infty$.

Because $\widehat{U}(c_1, \mathbf{c}) = \widehat{W}(c_1, \widehat{U}(\mathbf{c})) = \widehat{W}(c_1, f(U(\mathbf{c})))$ and also $\widehat{U}(c_1, \mathbf{c}) = f(W(c_1, U(\mathbf{c})))$, we have

$$f(W(c_1, U(\mathbf{c}))) = \widehat{W}(c_1, f(U(\mathbf{c})))$$

for all $c_1 \in C$ and $\mathbf{c} \in C^\infty$. □

Theorem 8.2 *Fix a continuous representation U of preference $\succsim$ over C^∞ which satisfies completeness, transitivity and continuity.*

Then, $\succsim$ satisfies Stationarity and Separability (b) if and only if there is a continuous function $u : C \to \mathbb{R}$ and a continuous and strongly increasing function $W : u(C) \times U(C^\infty) \to \mathbb{R}$ such that U has the form

$$U(c_1, \mathbf{c}) = W(u(c_1), U(\mathbf{c})).$$

Proof From Separability (a), which follows from Stationarity, and Separability (b), $\succsim$ is weakly separable over $C \times C^\infty$. Hence there exist functions $u : C \to \mathbb{R}$ and $U_2 : C^\infty \to \mathbb{R}$ and a monotone transformation $\widehat{W} : u(C) \times U_2(C^\infty) \to \mathbb{R}$ such that

$$U(c_1, \mathbf{c}) = \widehat{W}(u(c_1), U_2(\mathbf{c})).$$

Now fix an arbitrary $\widehat{c}_1 \in C$. Then, since U is a representation of $\succsim$,

$$U(\mathbf{c}) \geqq U(\mathbf{c}') \iff \mathbf{c} \succsim \mathbf{c}'$$

for any $\mathbf{c}, \mathbf{c}' \in C^\infty$.

On the other hand, since

$$\begin{aligned} & U_2(\mathbf{c}) \geqq U_2(\mathbf{c}') \\ \iff & \widehat{W}(u(\widehat{c}_1), U_2(\mathbf{c})) \geqq \widehat{W}(u(\widehat{c}_1), U_2(\mathbf{c}')) \\ \iff & (\widehat{c}_1, \mathbf{c}) \succsim (\widehat{c}_1, \mathbf{c}') \end{aligned}$$

for all $\mathbf{c}, \mathbf{c}' \in C^\infty$, Stationarity implies

$$U_2(\mathbf{c}) \geqq U_2(\mathbf{c}') \iff \mathbf{c} \succsim \mathbf{c}'.$$

Hence $U(\cdot)$ and $U_2(\cdot)$ represent the same preference $\succsim$ over C^∞, and there exists a monotone transformation $G : U(C^\infty) \to \mathbb{R}$ such that

$$U_2(\cdot) = G(U(\cdot)).$$

Now let $W : u(C) \times U(C^\infty) \to \mathbb{R}$ be such that

$$W(u, U_2) = \widehat{W}(u, G(U_2)).$$

□

Here we introduce a further separability axiom.

Axiom 8.6 (Separability II) For all $(c_1, c_2), (c_1', c_2') \in C^2$ and $\mathbf{c}, \mathbf{c}' \in C^\infty$

(c)

$$(c_1, c_2, \mathbf{c}) \succsim (c_1', c_2', \mathbf{c}) \iff (c_1, c_2, \mathbf{c}') \succsim (c_1', c_2', \mathbf{c}'),$$

(d)

$$(c_1, c_2, \mathbf{c}) \succsim (c_1', c_2, \mathbf{c}') \iff (c_1, c_2', \mathbf{c}) \succsim (c_1', c_2', \mathbf{c}').$$

Condition (c) states that preference over pairs of today's consumption and tomorrow's consumption is independent of what you receive from the day after tomorrow. Condition (d) states that preference over today's consumptions and sequences starting the day after tomorrow is independent of what you receive tomorrow.

Remark 8.1 Condition (d) is quite a technical one. It is known to be derived from Separability (a) (b) (c) and Lemma 8.2 according to Gorman [48], but we assume it directly since the proof is very technical.

Lemma 8.2 *Under Stationarity,*

$$(c_1, c_2, \mathbf{c}) \succsim (c_1, c_2, \mathbf{c}') \iff (c_1', c_2', \mathbf{c}) \succsim (c_1', c_2', \mathbf{c}') \qquad (*)$$

holds for all $(c_1, c_2), (c_1', c_2') \in C^2$ *and* $\mathbf{c}, \mathbf{c}' \in C^\infty$.

Proof It follows from

$$
\begin{aligned}
& (c_1, c_2, \mathbf{c}) \succsim (c_1, c_2, \mathbf{c}') \\
\text{(by Stationarity)} \iff & (c_2, \mathbf{c}) \succsim (c_2, \mathbf{c}') \\
\text{(by Separability (a) following from Stationarity)} \iff & (c_2', \mathbf{c}) \succsim (c_2', \mathbf{c}') \\
\text{(by Stationarity)} \iff & (c_1', c_2', \mathbf{c}) \succsim (c_1', c_2', \mathbf{c}').
\end{aligned}
$$

□

Lemma 8.3 *Under Stationarity and Separability (b),*

$$(c_1, c_2, \mathbf{c}) \succsim (c_1, c_2', \mathbf{c}) \iff (c_1', c_2, \mathbf{c}') \succsim (c_1', c_2', \mathbf{c}')$$

holds for all $(c_1, c_2), (c_1', c_2') \in C^2$ *and* $\mathbf{c}, \mathbf{c}' \in C^\infty$.

Proof It follows from

$$
\begin{aligned}
& (c_1, c_2, \mathbf{c}) \succsim (c_1, c_2', \mathbf{c}) \\
\text{(by Stationarity)} \iff & (c_2, \mathbf{c}) \succsim (c_2', \mathbf{c}) \\
\text{(by Separability (b))} \iff & (c_2, \mathbf{c}') \succsim (c_2', \mathbf{c}') \\
\text{(by Stationarity)} \iff & (c_1', c_2, \mathbf{c}') \succsim (c_1', c_2', \mathbf{c}').
\end{aligned}
$$

□

The theorem below characterizes the discounted utility representation.

Theorem 8.3 *Preference* $\succsim$ *over* C^∞ *satisfies Stationarity and Separability (b) (c) (d) as well as completeness, transitivity and continuity if and only if there exist a continuous function* $u : C \to \mathbb{R}$ *and a number* $\beta \in (0, 1)$ *such that* $\succsim$ *is represented in the form*

$$U(\mathbf{c}) = \sum_{t=1}^{\infty} u(c_t)\beta^{t-1}.$$

Also, u *is unique up to positive affine transformations and* β *is unique.*

Proof It suffices to show sufficiency of the axioms.

By Stationarity, Separability (b) (c) (d) and the above lemmata, $\succsim$ is strongly separable over $C \times C \times C^\infty$. Hence by Theorem 3.2 there exist functions u_1, u_2, U_3 such that

$$U(c_1, c_2, c_3, \mathbf{c}) = u_1(c_1) + u_2(c_2) + U_3(c_3, \mathbf{c}).$$

Without loss of generality, we can take some $c \in C$ such that $u_1(c) > 0$.

Fix some $\widehat{c}_1$. Then by Stationarity both $u_1(\widehat{c}_1)+u_2(c_2)+U_3(c_3,\mathbf{c})$ and $u_1(c_2)+u_2(c_3)+U_3(\mathbf{c})$ represent the same preference and the latter involves at least three additive components, by uniqueness up to positive transformation there exist $\beta > 0$ and γ such that

$$u_2(c_2) + U_3(c_3, \mathbf{c}) = \beta(u_1(c_2) + u_2(c_3) + U_3(\mathbf{c})) + \gamma$$

By doing this again, we have

$$\begin{aligned}
& U(c_1, c_2, c_3, c_4, \mathbf{c}) \\
&= u_1(c_1) + u_2(c_2) + U_3(c_3, c_4, \mathbf{c}) \\
&= u_1(c_1) + \beta(u_1(c_2) + u_2(c_3) + U_3(c_4, \mathbf{c})) + \gamma \\
&= u_1(c_1) + \beta u_1(c_2) + \beta(u_2(c_3) + U_3(c_4, \mathbf{c})) + \gamma \\
&= u_1(c_1) + \beta u_1(c_2) + \beta(\beta(u_1(c_3) + u_2(c_4) + U_3(\mathbf{c})) + \gamma) + \gamma \\
&= u_1(c_1) + \beta u_1(c_2) + \beta^2 u_1(c_3) + \beta^2(u_2(c_4) + U_3(\mathbf{c})) + \beta\gamma + \gamma.
\end{aligned}$$

and by repeating we obtain

$$\begin{aligned}
& U(c_1, c_2, \cdots, c_T, \mathbf{c}) \\
&= u_1(c_1) + \beta u_1(c_2) + \cdots + \beta^{T-2} u_1(c_{T-1}) + \beta^{T-2}(u_2(c_T) + U_3(\mathbf{c})) \\
&\quad +\gamma(1 + \beta + \cdots + \beta^{T-3}).
\end{aligned}$$

Suppose now $\beta \geqq 1$, for $c \in C$ with $u_1(c) > 0$ consider the constant sequence $(c, c, c \cdots)$, then we obtain

$$\begin{aligned}
U(c, c, c, \cdots) &= \lim_{T\to\infty} \left\{ \sum_{t=1}^{T-1} u_1(c)\beta^{t-1} + \beta^{T-2}(u_2(c) + U_3(c, c, c, \cdots)) \right. \\
&\quad \left. +\gamma(1 + \beta + \cdots + \beta^{T-3}) \right\} \\
&= \infty,
\end{aligned}$$

which contradicts to continuity of U with regard to the product metric over C^∞. Hence $0 < \beta < 1$.

Thus for all $\mathbf{c} \in C^\infty$ we obtain

$$\begin{aligned}
U(\mathbf{c}) = \lim_{T\to\infty} &\left\{ \sum_{t=1}^{T-2} u_1(c_t)\beta^{t-1} + \beta^{T-2}(u_2(c_T) + U_3(\mathbf{c})) \right. \\
&\left. +\gamma(1 + \beta + \cdots + \beta^{T-3}) \right\}
\end{aligned}$$

$$= \sum_{t=1}^{\infty} u_1(c_t)\beta^{t-1} + \frac{\gamma}{1-\beta}.$$

Now let $u_1 = u$.

Uniqueness up to positive affine transformations follows from the result on uniqueness of additively separable representation. □

8.1.1 Violation of Stationarity: Hyperbolic Discounting and Time Inconsistency

Consider the following two choice problems:

Problem 1: Receiving 1000 dollars after one year or 1100 dollars after one year and one week.

Problem 2: Receiving 1000 dollars right now or receiving 1100 dollars after one week.

Note that we are talking about consumption, that is, you cannot save the amount you receive and have to spend it on consumption on the day of receipt.

A typical experimental result is that subjects tend to choose the latter in Problem 1 but the former in Problem 2, but this cannot be explained by the discounted utility model. For, by assuming that consumption in other periods is unchanged and taking one period to be one week, the choice of the latter in Problem 1 means

$$u(1000)\beta^{50} < u(1100)\beta^{51},$$

which implies

$$u(1000) < u(1100)\beta,$$

meaning that one has to choose the latter in Problem 2 as well.

This suggests that realistic time discounting is rather hyperbolic than geometric/exponential. In other words, people tend to have larger discounting between today and tomorrow, saying "today is special," and the discounting tends to be flatter in later periods. Such a way of discounting is called hyperbolic discounting.

There is a tractable class of preferences which allows such a type of non-stationarity but still remains tractable. It is quasi-geometric discounting, in which preference at Period t is represented in the form

$$u(c_t) + \delta \sum_{\tau=t+1}^{\infty} u(c_\tau)\beta^{\tau-t-1}.$$

Here, δ is the discount factor between today and tomorrow. The case of $\delta = \beta$ corresponds to stationary discounting, while the case of $\delta \neq \beta$ describes the kind of impatience saying "today is special."

Of course, as stated already, there is a logical possibility that the decision maker exhibits an attitude saying "today is special, so I drink today and stop drinking tomorrow" and when tomorrow comes he indeed exhibits "the special day is over, so I never drink from today and onward." However, once you exhibit the kind of impatience saying "today is special" it is natural to assume that he exhibits the same kind of impatience tomorrow as well and onward.

He faces a dynamic inconsistency problem when he has the kind of impatience saying "today is special" each day. For, even he intends "today is special, so I drink today and stop drinking tomorrow," he will intend the same thing tomorrow.

When a decision maker faces such dynamic inconsistency, there will be roughly two modes of decision making. One is that he mistakenly believes that he can control his future selves and makes a plan, and then he fails to execute it, and he never learns and repeats the same mistake. This is called *naive* decision making. He is simply "stupid."

The other is that he accepts the fact that he cannot control his future selves and makes a plan by taking into account how his future selves behave. This is called *sophisticated* decision making. It is funny, because on one hand the decision maker lacks willpower in the sense of having the kind of impatience saying "today is special," and on the other hand he has an intellectual ability to understand the inconsistency precisely and to respond to it. Although such imbalance might be an aspect of reality, it looks funny.

The reality seems to be lying between the two, but to my knowledge there is not yet a well-accepted modeling of it.

Since modeling of naive decisions is pretty obvious, let me briefly illustrate a model of sophisticated decisions, following the approach by Strotz [95], Phelps and Pollak [86]. An individual earns income e every period and carries over assets k from the previous period, and decides how much to save for the next period. Let s denote the amount of saving, while the gross interest rate r is assumed to be constant over time. For later argument, we introduce a minimal saving amount b as a constraint.

This is a dynamic game being played against successive selves. Here we consider stationary Markov equilibrium, which is a stronger solution concept that subgame-perfect Nash equilibrium. Take it as given that future selves follow a saving function φ, then the lifetime utility from the viewpoint of the current self saving s is

$$u(k + e - s) + \delta V_\varphi(rs),$$

where V_φ is the continuation lifetime utility given by the recursive form

$$V_\varphi(k) = u(k + e - \varphi(k)) + \beta V_\varphi(r\varphi(k)).$$

When the current self maximizes the above lifetime utility and when the optimal choice is in fact what is given by φ, we say that the saving function φ is stationary

Markov equilibrium. That is, the saving function φ is such that

$$\varphi(k) \in \arg\max_{s \geq b} \left\{ u(k + e - s) + \delta V_{\varphi}(rs) \right\}.$$

Laibson [74] showed that making the minimal saving amount b larger may improve all the successive selves' lifetime utilities. This cannot happen in the standard models with dynamically consistent agents, since a constraint only narrows choice opportunities. That is, it shows that forced saving can work as a commitment device when an individual has a dynamically inconsistent sequence of preferences.

8.2 Risky Outcome Streams*

Here we follow Epstein [29] and extend the discounted utility theory to the domain of risky outcome streams.

The domain is $\Delta(C^{\infty})$, the set of lotteries over outcome streams in C^{∞}. We consider a preference relation $\succsim$ defined over $\Delta(C^{\infty})$.

First we impose axioms of the expected utility theory.

Axiom 8.7 (Expected Utility Theory) $\succsim$ satisfies the axioms of vNM expected utility theory.

Next we assume the natural extension of the Stationarity axiom. Let (c, l) denote a lottery in which the decision maker receives c for sure today and follows l from tomorrow and onward.

Axiom 8.8 (Stationarity) For all $c \in C$ and $l, l' \in \Delta(C^{\infty})$,

$$(c, l) \succsim (c, l') \iff l \succsim l'.$$

Stationarity includes the following separability condition, but let me state it separately as it makes sense by itself.

Axiom 8.9 (Risk Separability I) For all $c, c' \in C$ and $l, l' \in C^{\infty}$,

$$(c, l) \succsim (c, l') \iff (c', l) \succsim (c', l').$$

Theorem 8.4 *The following two statements are equivalent:*

(i) $\succsim$ *satisfies Expected Utility Theory and Stationarity.*
(ii) $\succsim$ *has an expected utility representation* $U : \Delta(C^{\infty}) \to \mathbb{R}$ *in the form*

$$U(l) = \int_{C^{\infty}} V(\mathbf{c}) l(d\mathbf{c})$$

such that its corresponding vNM index $V : C^\infty \to \mathbb{R}$ *has the form*

$$V(\mathbf{c}) = \sum_{t=1}^{\infty} u(c_t) \prod_{\tau=1}^{t-1} \beta(c_\tau)$$

for some $u : C \to \mathbb{R}$ *and* $\beta : C \to (0, 1)$.

Moreover, if another pair $(\widehat{u}, \widehat{\beta})$ *represents the same preference as* (u, β) *does in the above form, there exist numbers* A, B *with* $A > 0$ *such that*

$$\widehat{u}(\cdot) = Au(\cdot) + B(1 - \beta(\cdot)), \quad \widehat{\beta}(\cdot) = \beta(\cdot).$$

Proof It suffices to prove (ii) from (i).

Since $\succsim$ satisfies Expected Utility Theory, it has an expected utility representation

$$U(l) = \int_{C^\infty} V(\mathbf{c}) l(d\mathbf{c})$$

and U and V are unique up to positive affine transformations.

Pick any c_1, then the function over $\Delta(C^\infty)$ defined by

$$U(c_1, l) = \int_{C^\infty} V(c_1, \mathbf{c}) l(d\mathbf{c})$$

represents $\succsim$ because

$$\int_{C^\infty} V(c_1, \mathbf{c}) l(d\mathbf{c}) \geqq \int_{C^\infty} V(c_1, \mathbf{c}) m(d\mathbf{c}) \iff (c_1, l) \succsim (c_1, m)$$
$$\iff l \succsim m$$

follows from Stationarity, and we see that $V(c_1, \cdot)$ is a vNM index that forms an expected utility representation of $\succsim$.

From uniqueness of expected utility representation up to positive affine transformations, $V(c_1, \cdot)$ is a positive affine transformation of $V(\cdot)$ depending on c_1. Hence for some $u(c_1)$ and $\beta(c_1) > 0$,

$$V(c_1, \mathbf{c}) = u(c_1) + \beta(c_1) V(\mathbf{c}).s$$

By repeating this we obtain

$$V(c_1, \cdots, c_T, \mathbf{c}) = \sum_{t=1}^{T} u(c_t) \prod_{\tau=1}^{t-1} \beta(c_\tau) + \prod_{\tau=1}^{T} \beta(c_\tau) V(\mathbf{c}).$$

We show $\beta(c) < 1$ for all $c \in C$. Suppose there is $c \in C$ such that $\beta(c) \geqq 1$, then by considering a constant sequence $(c, c, c \cdots)$, where we assume $u(c) > 0$ without loss of generality, when $\beta(c) > 1$ we have

$$\begin{aligned} V(c, c, c, \cdots) &= \lim_{T \to \infty} u(c) \frac{\beta(c)^T - 1}{\beta(c) - 1} + \beta(c)^T V(c, c, c, \cdots) \\ &= \infty, \end{aligned}$$

which contradicts continuity with regard to product metric on C^∞. Likewise, when $\beta(c) = 1$ we have

$$\begin{aligned} V(c, c, c, \cdots) &= \lim_{T \to \infty} T u(c) + V(c, c, c, \cdots) \\ &= \infty, \end{aligned}$$

which leads to the same contradiction.

To show the second statement, suppose another pair $(\widehat{u}, \widehat{\beta})$ represents the same preference as (u, β) does in the above form. Then by cardinal equivalence there exist numbers A, B with $A > 0$ such that $\widehat{U}(\cdot) = AU(\cdot) + B$ and $\widehat{V}(\cdot) = AV(\cdot) + B$, where $\widehat{U}$ and $\widehat{V}$ are the associated expected utility function over $\Delta(C^\infty)$ and the associated vNM index over C^∞, respectively.

Because $\widehat{V}(c, y) = AV(c, y) + B$ and $\widehat{V}(c, y) = \widehat{u}(c) + \widehat{\beta}(c)\widehat{V}(y) = \widehat{u}(c) + \widehat{\beta}(c)\{AV(y) + B\}$, we have

$$V(c, y) = -\frac{B}{A} + \frac{1}{A}\widehat{u}(c) + \widehat{\beta}(c)V(y) + \frac{B}{A}\widehat{\beta}(c)$$

for all $c \in C$ and $y \in C^\infty$. Since $V(c, y) = u(c) + \beta(c)V(y)$, we have

$$u(c) + \beta(c)V(y) = -\frac{B}{A} + \frac{1}{A}\widehat{u}(c) + \widehat{\beta}(c)V(y) + \frac{B}{A}\widehat{\beta}(c)$$

for all $c \in C$ and $y \in C^\infty$, which is rewritten as

$$\{\beta(c) - \beta(c)\}\, V(y) = -\frac{B}{A} + \frac{1}{A}\widehat{u}(c) + \frac{B}{A}\widehat{\beta}(c) - u(c).$$

By applying this equality to different $y, y' \in C^\infty$, assuming $V(y) > V(y')$ without loss of generality, and taking the difference we obtain

$$\{\beta(c) - \beta(c)\}\left\{V(y) - V(y')\right\} = 0$$

for all $c \in C$. Thus we have $\widehat{\beta}(\cdot) = \beta(\cdot)$. Hence the left-hand side of the second-last formula is zero, which leads to $\widehat{u}(\cdot) = Au(\cdot) + B(1 - \beta(\cdot))$.

$$\widehat{u}(\cdot) = Au(\cdot) + B(1 - \beta(\cdot)), \quad \widehat{\beta}(\cdot) = \beta(\cdot).$$

□

Similarly to the deterministic case, we can show that the discount factor is a constant under an additional separability condition like below.

Axiom 8.10 (Risk Separability II) For all $l_{1,2}, l'_{1,2} \in \Delta(C^2)$ and for all $\mathbf{c}, \mathbf{c}' \in C^\infty$,

$$(l_{1,2}, \mathbf{c}) \succsim (l'_{1,2}, \mathbf{c}) \iff (l_{1,2}, \mathbf{c}') \succsim (l'_{1,2}, \mathbf{c}').$$

Theorem 8.5 *The following two statements are equivalent:*

(i) $\succsim$ *satisfies Expected Utility Theory, Stationarity and Risk Separability II.*
(ii) $\succsim$ *allows an expected utility representation* $U : \Delta(C^\infty) \to \mathbb{R}$ *in the form*

$$U(l) = \int_{C^\infty} V(\mathbf{c})l(d\mathbf{c})$$

such that the corresponding vNM index $V : C^\infty \to \mathbb{R}$ *has the form*

$$V(\mathbf{c}) = \sum_{t=1}^{\infty} u(c_t)\beta^{t-1}$$

for some $u : C \to \mathbb{R}$ *and* $\beta \in (0, 1)$.

Moreover, β *is unique and* u *is unique up to positive affine transformations.*

Proof When we restrict the preference relation $\succsim$ to the domain of deterministic sequences C^∞, following Remark 8.1, it is strongly separable over the partition $\{1\}, \{2\}, \{3, \cdots\}$ and $U(\cdot)$ is an *ordinal* representation of it by seeing that it is applied to the subdomain of degenerate lotteries. Hence $U(\cdot)$ is a *monotone transformation* of an additively separable representation. Thus we have

$$U(c_1, c_2, y) = F(\phi_1(c_1) + \phi_2(c_2) + \phi_3(y)).$$

Now fix an arbitrary $\overline{y}$. From Risk Separability II, $U(\cdot, \cdot, y)$ and $U(\cdot, \cdot, \overline{y})$ are *cardinally* equivalent given any y. Therefore, depending on y there are some $a(y)$

and $b(y) > 0$ such that

$$U(c_1, c_2, y) = a(y) + b(y)U(c_1, c_2, \overline{y})$$

holds.

Also, since $U(c_1, c_2, \cdot)$ and $\phi_3(\cdot)$ are *ordinally* equivalent, we have the form

$$U(c_1, c_2, y) = H(c_1, c_2, \phi_3(y)),$$

implying that the values of $a(y)$ and $b(y)$ depend only on $\phi_3(y)$. Thus we have the form

$$U(c_1, c_2, y) = \widehat{a}(\phi_3(y)) + \widehat{b}(\phi_3(y))U(c_1, c_2, \overline{y}).$$

Moreover, since $U(\cdot, \cdot, \overline{y})$ and $F(\phi_1(\cdot) + \phi_2(\cdot) + \phi_3(\overline{y}))$ are *ordinally* equivalent, for some monotone transformation G we have the form

$$U(c_1, c_2, \overline{y}) = G(F(\phi_1(c_2) + \phi_2(c_2) + \phi_3(\overline{y}))).$$

Here, by treating $\phi_3(\overline{y})$ as a constant, for some monotone transformation ψ we have the form

$$U(c_1, c_2, y) = \widehat{a}(\phi_3(y)) + \widehat{b}(\phi_3(y))\psi(\phi_1(c_1) + \phi_2(c_2)).$$

Summing up, we have the equality

$$F(\phi_1(c_1) + \phi_2(c_2) + \phi_3(y)) = \widehat{a}(\phi_3(y)) + \widehat{b}(\phi_3(y))\psi(\phi_1(c_1) + \phi_2(c_2))$$

and by letting $\phi_1(c_1) + \phi_2(c_2) = z$ and $\phi_3(y) = z'$ we obtain a functional equation

$$F(z + z') = \widehat{a}(z') + \widehat{b}(z')\psi(z).$$

There are two types of solutions to the above functional equation (see Corollary 1, Page 150 in Aczel [1]),

$$F(z) = \gamma e^{\alpha z} + \eta, \quad \gamma > 0,\ \alpha \neq 0,$$
$$F(z) = \gamma z + \eta, \quad \gamma > 0.$$

In the first case, assume $\gamma = \alpha = 1$ and $\eta = 0$ without loss of generality. Then

$$U(c_1, c_2, y) = \exp(\phi_1(c_1)) \exp(\phi_2(c_2) + \phi_3(y)),$$

but from Theorem 8.4 we already have the form

$$U(c_1, c_2, y) = u(c_1) + \beta(c_1)U(c_2, y)$$

and both can hold only when $u(c_1) = 0$ always hold, which is a contradiction to the previous theorem.

In the second case, assume $\gamma = 1$ and $\eta = 0$ without loss of generality. Then

$$U(c_1, c_2, y) = \phi_1(c_1) + \phi_2(c_2) + \phi_3(y)$$

and also it follows from Stationarity that

$$U(c_1, y) = \phi_1(c_1) + \rho U(y) + \sigma.$$

On the other hand, the previous theorem delivers the form

$$U(c_1, y) = u(c_1) + \beta(c_1)U(y)$$

and

$$\phi_1(c_1) + \rho U(y) + \sigma = u(c_1) + \beta(c_1)U(y)$$

for all (c_1, y). By applying this to any y, y' with $U(y) \neq U(y')$, we obtain $\beta(c_1) = \rho$ for all c_1 and we have shown that $\beta(\cdot)$ is a constant.

The second statement is now straightforward. □

8.3 Recursive Utility Under Risk (Kreps-Porteus/Epstein-Zin)*

Think of the following example as motivation.

Example 8.1 Consider choosing between the two below:

1. Flip a coin on the first day of your life. If it is heads you receive 100 dollar's worth of consumption every day, if it is tails you receive 1 dollar's worth of consumption every day.
2. Flip a coin each day. On each day, if it is heads you receive 100 dollar's worth of consumption, if it is tails you receive 1 dollar's worth of consumption that day.

In the standard expected discounted utility model

$$U(l) = E_l\left[\sum_{t=1}^{\infty} u(c_t)\beta^{t-1}\right],$$

where l denotes a lottery over consumption streams, by denoting the 100 dollar's worth of consumption by $\overline{c}$ and the 1 dollar's worth of consumption by $\underline{c}$, each option delivers the same expected lifetime utility $\frac{1}{2} \cdot \frac{v(\overline{c})}{1-\beta} + \frac{1}{2} \cdot \frac{v(\underline{c})}{1-\beta}$, but this is not intuitive.

Also, look at the standard parametric specification of the expected discounted utility model

$$E\left[\sum_{t=1}^{\infty} \frac{c_t^{1-\sigma}}{1-\sigma} \beta^{t-1}\right].$$

You see that attitudes toward two different issues are attributed to one parameter σ. When you restrict attention to streams with certainty, in the discounted utility evaluation

$$\sum_{t=1}^{\infty} \frac{c_t^{1-\sigma}}{1-\sigma} \beta^{t-1}$$

the parameter σ is the inverse of elasticity of intertemporal substitution. On the other hand, when you restrict attention to one-time gambles over constant consumption streams, in the expected utility evaluation

$$\frac{1}{1-\beta} E\left[\frac{c^{1-\sigma}}{1-\sigma}\right]$$

the parameter σ explains the degree of relative risk aversion. It is *empirically strange* that one parameter explains attitudes toward to different issues.

Here we follow Kreps and Porteus [73], Epstein and Zin [33], and Chew and Epstein [17], and consider a larger domain of multi-stage lotteries or probability trees, in order to distinguish between risk aversion and intertemporal substitution. We will drop indifference to timing, but maintain Dynamic Consistency, hence the standard dynamic programming argument is still applicable. The expected utility theory is assumed to hold for *timeless gambles*, but it is not imposed over gambles over time.

We will define the domain of infinite probability trees $\mathcal{D}$ as follows:

1. $\mathcal{D}_1 = \Delta(C)$ denotes the set of probability trees over one period (which may look too short to be called trees, though . . .).
2. $\mathcal{D}_2 = \Delta(C \times \mathcal{D}_1)$ denotes the set of probability trees over two periods.
3. Likewise, $\mathcal{D}_3 = \Delta(C \times \mathcal{D}_2) = \Delta(C \times \Delta(C \times \Delta(C)))$ denotes the set of probability trees over three periods.

   . . .

4. Inductively,

$$\mathcal{D}_t = \Delta(C \times \mathcal{D}_{t-1})$$

denotes the set of probability trees over t periods.

Since $\mathcal{D}_1$ is compact metric in the weak convergence topology, so is $\mathcal{D}_2$ and so is $\mathcal{D}_t$ for every t inductively. For moments, let us consider the set

$$\mathcal{D}^* = \prod_{t=1}^{\infty} \mathcal{D}_t.$$

This is because we cannot define a set like "$\mathcal{D}_\infty$" although we want to (one might be able to do so with a more advanced mathematical knowledge, but let us be content with the current one). Thus define an infinite probability tree as an infinite sequence of finite probability trees. Note that $\mathcal{D}^*$ is compact metric with regard to the product topology.

Such infinite probability trees must be defined as satisfying certain consistency conditions, since an arbitrary element of $\mathcal{D}^*$ may allow no consistency between its t-period entry and its $t-1$-period entry.

Define a mapping $G_1 : C \times \mathcal{D}_1 \to C$ by

$$G_1(c, p_1) = c$$

and define $H_1 : \mathcal{D}_2 \to \mathcal{D}_1$ as a mapping given by

$$H_1(p_2)(B_1) = p_2(G_1^{-1}(B_1))$$

for arbitrary Borel set $B_1 \in \mathcal{B}(C)$. Then the relation $p_1 = H_1(p_2)$ says that the probability distribution over first-period consumptions induced by the two-stage probability tree p_2 coincides with p_1.

Likewise, define a mapping $G_2 : C \times \mathcal{D}_2 \to C \times \mathcal{D}_1$ by

$$G_2(c, p_2) = (c, H_1(p_2))$$

and define $H_2 : \mathcal{D}_3 \to \mathcal{D}_2$ as a mapping given by

$$H_2(p_3)(B_2) = p_3(G_2^{-1}(B_2))$$

for arbitrary Borel set $B_2 \in \mathcal{B}(C \times \mathcal{D}_1)$. Then the relation $p_2 = H_2(p_3)$ says that the probability tree for the first two periods induced by p_3 coincides with p_2.

Inductively, define a mapping $G_t : C \times \mathcal{D}_t \to C \times \mathcal{D}_{t-1}$ by

$$G_t(c, p_t) = (c, H_{t-1}(p_t))$$

and define $H_t : \mathcal{D}_{t+1} \to \mathcal{D}_t$ as a mapping given by

$$H_t(p_{t+1})(B_t) = p_{t+1}(G_t^{-1}(B_t))$$

for arbitrary Borel set $B_t \in \mathcal{B}(C \times \mathcal{D}_{t-1})$. Then the relation $p_t = H_t(p_{t+1})$ says that the probability tree for the first t periods induced by p_{t+1} coincides with p_t.

Now define the domain of infinite probability trees by

$$\mathcal{D} = \left\{ p = (p_1, p_2, \cdots,) \in \mathcal{D}^* : p_t = H_t(p_{t+1}), \ \forall t \geq 1 \right\}.$$

Then such a domain is seen as equivalent to the domain of one-stage lotteries over pairs of current consumption and a probability tree starting in the next period, which is stated as in the theorem below.

Theorem 8.6

$$\mathcal{D} \simeq \Delta(C \times \mathcal{D}).$$

Proof It follows from Theorem 8.10, a more general one. □

Note that as a subspace of $\mathcal{D}^*$ the domain $\mathcal{D}$ is compact and metric.

Now we see

$$\mathcal{D} \supset \Delta(C^\infty) \supset C^\infty,$$

where $\Delta(C^\infty)$ is seen as the set of lotteries in which risk resolves in one period.

Consider the preference relation $\succsim$ defined over $\mathcal{D}$. We impose the following axioms.

Axiom 8.11 (Continuous Weak Order) $\succsim$ satisfies completeness, transitivity and continuity.

Axiom 8.12 (Mixture Independence for Static Risk) For all $p, p', p'' \in \mathcal{D}$ and $\lambda \in (0, 1)$,

$$p \succsim p' \implies (1-\lambda)p + \lambda p'' \succsim (1-\lambda)p' + \lambda p''.$$

This says we still impose mixture independence on timeless gambles, while we drop it for gambles across time.

The Stationarity axiom is the Dynamic Consistency in the current context. Note that $\delta(c, p)$ is the degenerate lottery that yields the pair of c and p for sure.

Axiom 8.13 (Stationarity) For all $p, p' \in \mathcal{D}$ and $c \in C$,

$$\delta(c, p) \succsim \delta(c, p') \iff p \succsim p'.$$

The above three axioms deliver the following recursive utility representation.

Theorem 8.7 *$\succsim$ satisfies Continuous Weak Order, Mixture Independence for Static Risk and Stationarity if and only if there exist functions $U : \mathcal{D} \to \mathbb{R}$ and $W : C \times U(\mathcal{D}) \to \mathbb{R}$ such that U represents $\succsim$ and has the recursive form*

$$U(p) = \int_{C\times\mathcal{D}} W\left(c', U(p')\right) dp(c', p')$$

which holds for all $p \in \mathcal{D}$.

Moreover, if $(\widehat{U}, \widehat{W})$ forms a representation of $\succsim$ in the above way then there exist $\alpha > 0$ and $\beta \in \mathbb{R}$ such that

$$\widehat{U} = \alpha U + \beta$$

and it holds for all $\widehat{V} \in \widehat{U}(\mathcal{D})$ that

$$\widehat{W}(c, \widehat{V}) = \alpha W\left(c, \frac{\widehat{V} - \beta}{\alpha}\right) + \beta.$$

Proof It suffices to show sufficiency of the axioms.

Since $\succsim$ satisfies the expected utility theory over $\Delta(C \times \mathcal{D})$, there is a function $V : C \times \mathcal{D} \to \mathbb{R}$ such that $\succsim$ is represented in the expected utility form

$$U(p) = \int_{C\times\mathcal{D}} V\left(c', p'\right) dp(c', p').$$

Now fix an arbitrary $c \in C$. Then by Stationarity,

$$\begin{aligned} V(c, p) \geq V(c, p') &\iff \delta(c, p) \succsim \delta(c, p') \\ &\iff p \succsim p' \\ &\iff U(p) \succsim U(p'), \end{aligned}$$

which means that $U(\cdot)$ and $V(c, \cdot)$ are *ordinally* equivalent. Hence there is a monotone transformation $W(c, \cdot) : U(\mathcal{D}) \to \mathbb{R}$, which depends on c, such that

$$V(c, p) = W(c, U(p))$$

holds.

The uniqueness result follows from uniqueness of U and V up to positive affine transformations. □

The general class as above reduces to the standard expected utility model under the following additional axiom stating that the decision maker is indifferent to time of resolution of risk.

Axiom 8.14 (Timing Indifference) For all $c \in C$, $p, p' \in \mathcal{D}$ and $\lambda \in [0, 1]$,

$$\lambda\delta(c, p) + (1 - \lambda)\delta(c, p') \sim \delta(c, \lambda p + (1 - \lambda)p').$$

Theorem 8.8 *$\succsim$ satisfies Continuous Weak Order, Mixture Independence for Static Risk, Stationarity and Timing Indifference if and only if there exists a function $U : \mathcal{D} \to \mathbb{R}$ that represents $\succsim$ in the expected utility form*

$$U(p) = = \int_{C^\infty} V(\mathbf{c}) l_p(d\mathbf{c})$$

such that its corresponding vNM index $V : C^\infty \to \mathbb{R}$ has the form

$$V(\mathbf{c}) = \sum_{t=0}^{\infty} u(c_t) \prod_{\tau=0}^{t-1} \beta(c_\tau)$$

with $u : C \to \mathbb{R}$ and $\beta : C \to (0, 1)$, where $l_p \in \Delta(C^\infty)$ denotes the one-shot lottery induced by $p \in \mathcal{D}$.

Proof From

$$\begin{aligned} U(\lambda\delta(c, p) + (1 - \lambda)\delta(c, p')) &= \lambda V(c, p) + (1 - \lambda)V(c, p') \\ &= \lambda W(c, U(p)) + (1 - \lambda)W(c, U(p')) \end{aligned}$$

and

$$\begin{aligned} U(\delta(c, \lambda p + (1 - \lambda)p')) &= V(c, \lambda p + (1 - \lambda)p') \\ &= W(c, U(\lambda p + (1 - \lambda)p')) \\ &= W(c, \lambda U(p) + (1 - \lambda)U(p')) \end{aligned}$$

we obtain

$$W(c, \lambda U(p) + (1 - \lambda)U(p')) = \lambda W(c, U(p)) + (1 - \lambda)W(c, U(p'))$$

and we see that $W(c, U)$ is mixture-linear in the second argument. Hence we obtain the form

$$W(c, U) = u(c) + \beta(c)U$$

and it follows that $\beta(c) > 0$ from monotonicity of W in the second argument. It also follows that $\beta(c) < 1$ from the previous argument. □

8.3.1 Separation of Elasticity of Intertemporal Substitution and Risk Aversion

Let us verify that the above class of recursive utilities allows separation of elasticity of intertemporal substitution and risk aversion.

We define comparative risk aversion as follows, regarding timeless risks.

Definition 8.1 Say that $\succsim^a$ is more risk-averse than $\succsim^b$ if for all $l \in \Delta(C^\infty)$ and $y \in C^\infty$,

$$l \succsim^a y \implies l \succsim^b y$$

and

$$l \succ^a y \implies l \succ^b y.$$

Note that comparison of risk attitudes makes sense only when preferences over deterministic consumption sequences coincide.

Theorem 8.9 *$\succsim^a$ is more risk-averse than $\succsim^b$ if and only if there is an increasing and concave function $\phi : U^b(C^\infty) \to U^a(C^\infty)$ such that for all $y \in C^\infty$*

$$U^a(y) = \phi(U^b(y))$$

and for all $c \in C$ and $V \in U^b(C^\infty)$,

$$W^a(c, \phi(V)) = \phi(W^b(c, V)).$$

Proof Because $\succsim^a$ and $\succsim^b$ coincide over C^∞ there is a monotone function $\phi : U^b(C^\infty) \to U^a(C^\infty)$ such that

$$U^a(y) = \phi(U^b(y)).$$

Also, since $U^a(y) = \phi(U^b(y)) = \phi(W^b(c, U^b(y)))$ and $U^a(c, y) = W^a(c, U^a(y)) = W^a(c, \phi(U^b(y)))$, we obtain

$$W^a(c, \phi(U^b(y))) = \phi(W^b(c, U^b(y))).$$

To show that ϕ is concave, consider a one-shot lottery $l = \lambda\delta(y') + (1-\lambda)\delta(y'')$. Then

$$\begin{aligned} U^a(l) &= \lambda U^a(y') + (1-\lambda)U^a(y'') \\ &= \lambda\phi(U^b(y')) + (1-\lambda)\phi(U^b(y'')), \\ U^b(l) &= \lambda U^b(y') + (1-\lambda)U^b(y''). \end{aligned}$$

Note that $U^a(l) \geq U^a(y)$ is equivalent to $\lambda U^a(y')+(1-\lambda)U^a(y'') \geq U^a(y)$, that is, $\lambda\phi(U^b(y')) + (1-\lambda)\phi(U^b(y'')) \geq \phi(U^b(y))$. Note also that $U^b(l) \geq U^b(y)$ is equivalent to $\lambda U^b(y') + (1-\lambda)U^b(y'') \geq U^b(y)$, that is, $\phi(\lambda U^b(y') + (1-\lambda)U^b(y'')) \geq \phi(U^b(y))$. Then the required conditions met only when

$$\phi(\lambda U^b(y') + (1-\lambda)U^b(y'')) \geq \lambda\phi(U^b(y')) + (1-\lambda)\phi(U^b(y'')),$$

which establishes concavity of ϕ.

The converse is straightforward. □

8.4 Streams Under Subjective Uncertainty*

This section extends the theory of recursive utility to the setting of subjective uncertainty, following Hayashi [57]. The purpose is to enable the three-way separation of elasticity of intertemporal substitution, risk aversion and ambiguity aversion.

8.4.1 The Choice Domain

Let C denote the set of consumptions per period, which is assumed to be compact metric. Let S denote the set of states per period, which is assumed to be finite. Given a set Y, the set of functions from S to Y is denoted by Y^S.

We define the domain $\mathcal{H}$ as follows:

1. $\mathcal{H}_1 = \Delta(C)^S$ denotes the set of uncertain streams over one period (which might be too short to be called streams ...), which are lottery-acts as in the static model. It is compact metric.
2. $\mathcal{H}_2 = \Delta(C \times \mathcal{H}_1)^S$ denotes the set of uncertain streams over two periods, which map each one-step-ahead realization of uncertainty into a lottery over pairs of one-period consumption and an act in $\mathcal{H}_1$. It is compact metric.
3. Likewise, $\mathcal{H}_3 = \Delta(C \times \mathcal{H}_2)^S = \Delta(C \times \Delta(C \times \Delta(C)^S)^S)^S$ denotes the set of uncertain streams over three periods, which map each one-step-ahead realization of uncertainty into a lottery over pairs of one-period consumption and an act in $\mathcal{H}_2$. It is compact metric.

   ...

4. Inductively,

$$\mathcal{H}_t = \Delta(C \times \mathcal{H}_{t-1})^S$$

denotes the set of uncertain streams over t periods, which map each one-step-ahead realization of uncertainty into a lottery over pairs of one-period consumption and an act in $\mathcal{H}_{t-1}$. It is compact metric.

For moments, let us consider the set

$$\mathcal{H}^* = \prod_{t=1}^{\infty} \mathcal{H}_t.$$

Again, this is because we cannot define a set like "$\mathcal{H}_\infty$" although we want to (one might be able to do so with a more advanced mathematical knowledge, but let us be content with the current one). Thus define an infinite tree as an infinite sequence of finite trees. Note that $\mathcal{H}^*$ is compact metric with regard to the product topology.

As in the construction of the domain of infinite probability trees, the definition of $\mathcal{H}^*$ imposes no consistency between the t-period entry and its $t-1$-period entry. This requires us to go through the following argument.

Define a mapping $G_1 : C \times \mathcal{H}_1 \to C$ by

$$G_1(c, h_1) = c$$

and define $H_1 : \mathcal{H}_2 \to \mathcal{H}_1$ as a mapping given by

$$H_1(h_2)(s)(B_1) = h_2(s)(G_1^{-1}(B_1))$$

for each $s \in S$ and arbitrary Borel set $B_1 \in \mathcal{B}(C)$. Here the relation $h_1 = H_1(h_2)$ states that the one-period state-contingent consumption lottery for the first period induced by the two-period uncertain stream h_2 coincides with h_1.

Likewise, define a mapping $G_2 : C \times \mathcal{H}_2 \to C \times \mathcal{H}_1$ by

$$G_2(c, h_2) = (c, H_1(h_2))$$

and define $H_2 : \mathcal{H}_3 \to \mathcal{H}_2$ as a mapping given by

$$H_2(h_3)(s)(B_2) = h_3(s)(G_2^{-1}(B_2))$$

for each $s \in S$ and arbitrary Borel set $B_2 \in \mathcal{B}(C \times \mathcal{H}_1)$. Here the relation $h_2 = H_2(h_3)$ states that the two-period uncertain consumption streams for the first two periods induced by the three-period uncertain stream h_3 coincides with h_2.

Inductively, define a mapping $G_t : C \times \mathcal{H}_t \to C \times \mathcal{H}_{t-1}$ by

$$G_t(c, h_t) = (c, H_{t-1}(h_t))$$

and define $H_t : \mathcal{H}_{t+1} \to \mathcal{H}_t$ as a mapping given by

$$H_t(h_{t+1})(s)(B_t) = h_{t+1}(s)(G_t^{-1}(B_t))$$

for each $s \in S$ and arbitrary Borel set $B_t \in \mathcal{B}(C \times \mathcal{H}_{t-1})$. Here the relation $h_t = H_t(h_{t+1})$ states that the $t-1$ period uncertain consumption streams for the first $t-1$ periods induced by the $t+1$ period uncertain stream h_{t+1} coincide with h_t.

Now define the domain of infinitely long uncertain consumption streams by

$$\mathcal{H} = \left\{h = (h_1, h_2, \cdots,) \in \mathcal{H}^* : h_t = H_t(h_{t+1}),\ \forall t \geq 1\right\}.$$

Such an infinitely long state-contingent consumption stream is viewed as a one-period random variable which maps each realization of one-step-ahead uncertainty into a lottery over pairs of current consumptions and infinite uncertain streams starting tomorrow. This is what the theorem below says.

Theorem 8.10

$$\mathcal{H} \simeq \Delta(C \times \mathcal{H})^S.$$

The proof consists of several lemmata.

Given $h \in \mathcal{H}$ and $s \in S$, the sequence defined by $(h_1(s), h_2(s), h_3(s), \cdots) \in \prod_{t=1}^{\infty} \Delta(C \times \mathcal{H}_{t-1})$ is viewed as a sequence of constant mappings satisfying

$$\begin{aligned}
h_1(s)(s') &= h_1(s) \in \Delta(C) \\
h_2(s)(s') &= h_2(s) \in \Delta(C \times \mathcal{H}_1) \\
&\vdots \\
h_t(s)(s') &= h_t(s) \in \Delta(C \times \mathcal{H}_{t-1}) \\
&\vdots
\end{aligned}$$

for all $s' \in S$. For such objects, we obtain the following two lemmata.

Lemma 8.4 *For all $h \in \mathcal{H}$ and $s \in S$, the sequence $(h_1(s), h_2(s), h_3(s), \cdots) \in \prod_{t=1}^{\infty} \Delta(C \times \mathcal{H}_{t-1})$ satisfies $h_t(s) = H_t(h_{t+1}(s))$ for all $t \geq 1$.*

Proof For all $s' \in S$ and $B_t \in \mathcal{B}(C \times \mathcal{H}_{t-1})$, the claim follows from

$$\begin{aligned}
& h_t(s)(s')(B_t) \\
\text{(From the definition of } h_t(s) \text{ as a constant mapping)} &= h_t(s)(B_t) \\
\text{(From } h \in \mathcal{H}) &= H_t(h_{t+1})(s)(B_t) \\
\text{(From the definition of } H_t) &= h_{t+1}(s)(G_t^{-1}(B_t)) \\
\text{(From the definition of } h_{t+1}(s) \text{ as a constant mapping)} &= h_{t+1}(s)(s')(G_t^{-1}(B_t)) \\
\text{(From the definition of } H_t) &= H_t(h_{t+1}(s))(s')(B_t).
\end{aligned}$$

□

Lemma 8.5 *For all $t \geq 1$, if $h_t \in \mathcal{H}_t$ and $h_{t+1} \in \mathcal{H}_{t+1}$ satisfy*

$$h_t(s) = H_t(h_{t+1}(s))$$

for all $s \in S$, then

$$h_t = H_t(h_{t+1}).$$

Proof Fix an arbitrary $s \in S$. Then, for all $B_t \in \mathcal{B}(C \times \mathcal{H}_{t-1})$ and $s' \in S$,

$$\begin{aligned} & H_t(h_{t+1})(s)(B_t) \\ \text{(From the definition of } H_t) &= h_{t+1}(s)(G_t^{-1}(B_t)) \\ \text{(From the definition of } h_{t+1}(s) \text{ as a constant mapping)} &= h_{t+1}(s)(s')(G_t^{-1}(B_t)) \\ \text{(From the definition of } H_t) &= H_t(h_{t+1}(s))(s')(B_t) \\ \text{(From the previous lemma)} &= h_t(s)(s')(B_t) \\ \text{(From the definition of } h_t(s) \text{ as a constant mapping)} &= h_t(s)(B_t). \end{aligned}$$

Since $s \in S$ was arbitrary, we obtain $h_t = H_t(h_{t+1})$. □

Let

$$Z = \left\{ z = (z_1, z_2, \cdots) \in \prod_{t=1}^{\infty} \Delta(C \times \mathcal{H}_{t-1}) : z_t = H_t(z_{t+1}),\ \forall t \geq 1 \right\}.$$

Note that $\mathcal{H}_0$ is an empty set.

Then we have

Lemma 8.6

$$\mathcal{H} \simeq Z^S$$

Proof Define a mapping $\varphi : \mathcal{H} \to Z^S$ by

$$\varphi(h)(s) = (h_1(s), h_2(s), h_3(s), \cdots)$$

for each $s \in S$. Then, from the previous lemma we obtain $\varphi(h) \in Z^S$ and we see that φ is well-defined.

One-to-one: Suppose $\varphi(h) = \varphi(h')$, then it means

$$(h_1(s), h_2(s), h_3(s), \cdots) = (h'_1(s), h'_2(s), h'_3(s), \cdots)$$

holds for every $s \in S$, hence we obtain $h = h'$.

Onto: Pick any $\widetilde{h} \in Z^S$, then it gives

$$\widetilde{h}(s) = (\widetilde{h}_1(s), \widetilde{h}_2(s), \widetilde{h}_3(s), \cdots) \in \prod_{t=1}^{\infty} \Delta(C \times \mathcal{H}_{t-1})$$

for each $s \in S$. Thus, take $(h_1, h_2, \cdots,) \in \mathcal{H}^*$ such that $h_t(s) = \widetilde{h}_t(s)$ is met for each $t \geq 1$ and $s \in S$, then $\varphi^{-1}(\widetilde{h}) = (h_1, h_2, \cdots)$. From the previous lemma we see that such $(h_1, h_2, \cdots)$ satisfies $h_t = H_t(h_{t+1})$ for all $t \geq 1$, hence it is in $\mathcal{H}$.

Continuity: Suppose $h^\nu \to h$. From the property of product topology over $\mathcal{H}$, $h_t^\nu(s) \to h_t(s)$ for all $t \geq 1$ and $s \in S$. From the property of product topology over Z, $(h_1^\nu(s), h_2^\nu(s), h_3^\nu(s), \cdots) \to (h_1(s), h_2(s), h_3(s), \cdots)$ for each fixed $s \in S$. Hence in the product topology for Z^S, $\varphi(h^\nu) \to \varphi(h)$. □

Therefore, once we establish

$$Z \simeq \Delta(C \times \mathcal{H}),$$

we obtain the theorem because

$$X \simeq Y \implies X^S \simeq Y^S$$

when S is finite.

Now let

$$M^* = \left\{ (m_1, m_2, \cdots) \in \prod_{t=1}^{\infty} \Delta\left(C \times \prod_{\tau=1}^{t-1} \mathcal{H}_\tau\right) : \right.$$
$$\left. mrg_{C \times \prod_{\tau=1}^{t-1} \mathcal{H}_\tau} m_{t+1} = m_t, \quad \forall t \geq 1 \right\}$$

Then from the Kolmogorov extension theorem (Theorem 15.26 in Aliprantis and Border [3]) we have the following.

Lemma 8.7 *For all $(m_1, m_2, \cdots) \in M^*$ there is unique $q \in \Delta(C \times \mathcal{H}^*)$ such that*

$$mrg_{C \times \prod_{\tau=1}^{t-1} \mathcal{H}_\tau} q = m_t, \quad \forall t \geq 1$$

holds.

When we denote the mapping given in this way by ψ, it is a homeomorphism and $\psi(\{m_t\})$ gives the Kolmogorov extension of $\{m_t\}$.

For arbitrary $t \geq 0$, define

$$\mathcal{H}^t = \left\{ (h_1, \cdots, h_t) \in \prod_{\tau=1}^{t} \mathcal{H}_\tau : h_\tau = H_\tau(h_{\tau+1}),\ \ \forall \tau = 1, \cdots, t-1 \right\}.$$

Note that $\mathcal{H}^1 = \mathcal{H}_1$ and $\mathcal{H}^0$ is empty.

Let

$$M = \{(m_1, m_2, \cdots) \in M^* : m_t(C \times \mathcal{H}^{t-1}) = 1,\ \forall t \geq 1\}$$

then we obtain the following.

Lemma 8.8 $\psi(M) = \Delta(C \times \mathcal{H})$. *Hence*

$$M \simeq \Delta(C \times \mathcal{H}).$$

Proof First we show $\psi(M) \subset \Delta(C \times \mathcal{H})$. Pick $q \in \Delta(C \times \mathcal{H}^*)$ such that there is $\{m_t\} \in M$ giving $q = \psi(\{m_t\})$.

For each $t \geq 1$, define

$$\Gamma_t = C \times \mathcal{H}^t \times \prod_{\tau=t+1}^{\infty} \mathcal{H}_t.$$

Then we obtain

$$C \times \mathcal{H} \subset \Gamma_t \subset C \times \mathcal{H}^*.$$

The sequence $\{\Gamma_t\}$ is decreasing and $\bigcap_{t=1}^{\infty} \Gamma_t = C \times \mathcal{H}$.

Since q is the Kolmogorov extension of $\{m_t\}$ it satisfies

$$q(\Gamma_t) = m_t(C \times \mathcal{H}^t) = 1$$

for every $t \geq 1$. Hence we obtain

$$q(C \times \mathcal{H}) = q\left(\bigcap_{t=1}^{\infty} \Gamma_t\right) = \lim q(\Gamma_t) = 1.$$

Next we show $\psi(M) \supset \Delta(C \times \mathcal{H})$. Pick any $q \in \Delta(C \times \mathcal{H})$ then it must be that $q(C \times \mathcal{H}) = 1$. Now define $\{m_t\}$ as the sequence of marginal distributions given by

$$m_t = mrg_{C \times \prod_{\tau=1}^{t-1}} q$$

for each $t \geq 1$. Then, from $m_t(C \times \mathcal{H}^t) = q(\Gamma_t)$ and $\Gamma_t \supset C \times \mathcal{H}$, $q(\Gamma_t) \geq 1$, hence we obtain $m_t(C \times \mathcal{H}^t) = 1$.

Since q is associated with such $\{m_t\} \in M$, we obtain $q = \psi(\{m_t\})$. □

Lemma 8.9 *For all* $(z_t) \in Z$ *there is unique* $(m_t) \in M$ *such that*

$$mrg_{C \times \mathcal{H}_{t-1}} m_t = z_t$$

for all $t \geq 1$.

Denote the mapping defined in such way by ϕ, *then it is a homeomorphism and*

$$Z \simeq M.$$

Proof Define a sequence of mappings $\{\zeta_t\}$, such that $\zeta_t : C \times \mathcal{H}_{t-1} \to C \times \prod_{\tau=1}^{t-1} \mathcal{H}_\tau$ is given by

$$\zeta_t(c, h_{t-1}) = (c, \widehat{h}_1, \cdots, \widehat{h}_{t-1})$$

for each $t \geq 1$, where $\widehat{h}_{t-1} = h_{t-1}$ and $\widehat{h}_\tau = H_\tau(\widehat{h}_{\tau+1})$ for each $\tau = 1, 2, \cdots, t-2$. Note that ζ_1, ζ_2 are identity mappings.

By definition, ζ_t is one-to-one and $\zeta_t(C \times \mathcal{H}_{t-1}) = C \times \mathcal{H}^{t-1}$.

Next, for each $t \geq 1$, define a mapping $\xi_t : C \times \prod_{\tau=1}^{t-1} \mathcal{H}_\tau \to C \times \mathcal{H}_{t-1}$ as the projection

$$\xi_t(c, h_1, \cdots, h_{t-1}) = (c, h_{t-1})$$

Then, over $C \times \mathcal{H}^{t-1}$, $\xi_t = \zeta_t^{-1}$, and we obtain $\xi_t(C \times \mathcal{H}^{t-1}) = \zeta_t^{-1}(C \times \mathcal{H}^{t-1}) = C \times \mathcal{H}_{t-1}$.

Now, given any $\{z_t\} \in Z$, define the corresponding $\{m_t\} \in M$ as the one given by

$$m_t(A_t) = z_t(\xi_t(A_t))$$

for arbitrary Borel set $A_t \in \mathcal{B}(C \times \prod_{\tau=1}^{t-1} \mathcal{H}_\tau)$.

Then

$$m_t(C \times \mathcal{H}^{t-1}) = z_t(\xi_t(C \times \mathcal{H}^{t-1})) = z_t(C \times \mathcal{H}_{t-1}) = 1.$$

Also, by definition,

$$mrg_{C \times \mathcal{H}_{t-1}} m_t = z_t.$$

for every $t \geq 1$.

We define the mapping $\phi : Z \to M$ in the above way. We show that it is a homeomorphism.

One-to-one: Suppose $\phi(\{z_t\}) = \phi(\{z'_t\})$. Then it means

$$z_t(\xi_t(A_t)) = z'_t(\xi_t(A_t))$$

holds for all $t \geq 1$ and Borel set $A_t \in \mathcal{B}(C \times \prod_{\tau=1}^{t-1} \mathcal{H}_\tau)$. Since ξ_t is projection, the above means $z_t(B_t) = z'_t(B_t)$ holds for arbitrary Borel set $B_t \in \mathcal{B}(C \times \mathcal{H}_{t-1})$. Thus we obtain $\{z_t\} = \{z'_t\}$.

Onto: When $\{m_t\} \in M$ is given, we can take $\{z_t\} \in Z$ such that

$$z_t(B_t) = m_t(\xi_t^{-1}(B_t))$$

for arbitrary Borel set $B_t \in \mathcal{B}(C \times \mathcal{H}_{t-1})$.

Continuity: It follows from continuity of projection ξ_t for every $t \geq 1$ and the property of product topology. □

From the above, we obtain $Z \simeq M \simeq \Delta(C \times \mathcal{H})$, which yields $Z \simeq \Delta(C \times \mathcal{H})$, and we obtain

$$\mathcal{H} \simeq Z^S \simeq \Delta(C \times \mathcal{H})^S.$$

The recursive domain of lottery-acts $\mathcal{H}$ includes the domain of probability trees and further narrower subdomains. That is,

$$\mathcal{H} \supset \mathcal{D} \supset \Delta(C^\infty) \supset C^\infty.$$

In particular, we will often consider a subdomain in which subjective uncertainty resolves in one period and is relevant only between today and tomorrow, which is given by

$$\mathcal{H}_{+1} = \mathcal{D}^S.$$

8.4.2 The Axioms

For Period t, let $s^{t-1} \in S^{t-1}$ denote a history up to the previous period. When history s^{t-1} is given, the decision maker has preference $\succsim_{s^{t-1}}$ that is defined over $\Delta(C \times \mathcal{H})$, the set of lotteries over pairs of current consumption and uncertain process. Thus we consider a process of such preferences, $\{\succsim_{s^{t-1}}\}$.

We impose the following axioms. The first four will be self-explanatory.

Axiom 8.15 (Continuous Weak Order) For all t and $s^{t-1} \in S^{t-1}$, the conditional preference $\succsim_{s^{t-1}}$ satisfies completeness, transitivity and continuity.

Axiom 8.16 (Current Consumption Separability) For all t and $s^{t-1} \in S^{t-1}$, for all $h, h'\mathcal{H}$ and $c, c' \in C$,

$$\delta(c, h) \succsim_{s^{t-1}} \delta(c, h') \iff \delta(c', h) \succsim_{s^{t-1}} \delta(c', h').$$

Axiom 8.17 (Mixture Independence for Static Risks) For all t and $s^{t-1} \in S^{t-1}$, for all $p, p', p'' \in \Delta(C \times \mathcal{H})$ and $\lambda \in (0, 1)$,

$$p \succsim_{s^{t-1}} p' \implies (1-\lambda)p + \lambda p'' \succsim_{s^{t-1}} (1-\lambda)p' + \lambda p''.$$

Axiom 8.18 (History Independence of Risk Preference) For all $t, \widetilde{t}, s^{t-1} \in S^{t-1}$ and $\widetilde{s}^{\widetilde{t}-1} \in S^{\widetilde{t}-1}$,

$$p \succsim_{s^{t-1}} p' \iff p \succsim_{\widetilde{s}^{\widetilde{t}-1}} p'$$

holds for all $p, p' \in \mathcal{D}$,

Next we impose two axioms due to Gilboa and Schmeidler over the subdomain of lottery-acts with which uncertainty is resolved in one period.

Axiom 8.19 (Certainty Independence) For all t and $s^{t-1} \in S^{t-1}$, for all $c \in C$,

$$\begin{aligned} &\delta(c, h_{+1}) \succsim_{s^{t-1}} \delta(c, \widetilde{h}_{+1}) \\ &\implies \delta(c, (1-\lambda)h_{+1} + \lambda d) \succsim_{s^{t-1}} \delta(c, (1-\lambda)\widetilde{h}_{+1} + \lambda d) \end{aligned}$$

holds for all $h_{+1}, \widetilde{h}_{+1} \in \mathcal{H}_{+1}$, $d \in \mathcal{D}$ and $\lambda \in [0, 1]$.

Axiom 8.20 (Uncertainty Aversion) For all t and $s^{t-1} \in S^{t-1}$, for all $c \in C$,

$$\begin{aligned} &\delta(c, h_{+1}) \sim_{s^{t-1}} \delta(c, \widetilde{h}_{+1}) \\ &\implies \delta(c, (1-\lambda)h_{+1} + \lambda\widetilde{h}_{+1}) \succsim_{s^{t-1}} \delta(c, h_{+1}) \end{aligned}$$

holds for all $h_{+1}, \widetilde{h}_{+1} \in \mathcal{H}_{+1}$ and $\lambda \in [0, 1]$.

The last axiom is Dynamic Consistency.

Axiom 8.21 (Dynamic Consistency) For all t and $s^{t-1} \in S^{t-1}$, for all $c \in C$,

$$h(s) \succsim_{s^{t-1},s} \widetilde{h}(s),\ \forall s \in S \implies \delta(c, h) \succsim_{s^{t-1}} \delta(c, \widetilde{h})$$

holds for all $h, \widetilde{h} \in \mathcal{H}$.

8.4.3 The Representation Theorem

Theorem 8.11 *The preference process* $\{\succsim_{s^{t-1}}\}_{s^{t-1}\in S^{t-1},t\geq 1}$ *satisfies the above six axioms if and only if there exists* $(\{U_{s^{t-1}}, M_{s^{t-1}}\}, W)$ *such that for all* t *and* s^{t-1} *the function* $U_{s^{t-1}}$ *represents* $\succsim_{s^{t-1}}$ *in the recursive form*

$$U_{s^{t-1}}(p) = \int_{C\times\mathcal{H}} W\left(c, \min_{\mu\in M_{s^{t-1}}} \sum_{s\in S} U_{s^{t-1},s}(h(s))\mu(s)\right) dp(c,h).$$

Moreover, another $(\{\widehat{U}_{s^{t-1}}, \widehat{M}_{s^{t-1}}\}, \widehat{W})$ *represents* $\{\succsim_{s^{t-1}}\}_{s^{t-1}\in S^{t-1},t\geq 1}$ *in the above way if and only if* $A > 0$ *and* $B \in \mathbb{R}$ *such that*

$$\widehat{U}_{s^{t-1}} = AU_{s^{t-1}} + B$$

$$\widehat{M}_{s^{t-1}} = M_{s^{t-1}}$$

holds for all t *and* s^{t-1}, *and*

$$\widehat{W}(c, \widehat{V}) = \alpha W\left(c, \frac{\widehat{V} - \beta}{A}\right) + B$$

holds for all $\widehat{V} \in \widehat{U}(D)$.

Proof Because $\succsim_{s^{t-1}}$ satisfies the condition for the vNM expected utility theory over $\Delta(C \times \mathcal{H})$, there is $u_{s^{t-1}} : C \times \mathcal{H}$ such that

$$U_{s^{t-1}}(p) = \int_{C\times\mathcal{H}} u_{s^{t-1}}(c,h)dp(c,h)$$

holds.

From Current Consumption Separability, $u_{s^{t-1}}$ has the form

$$u_{s^{t-1}}(c,h) = W_{s^{t-1}}(c, u_{s^{t-1}}(\widehat{c}, h))$$

by fixing some arbitrary $\widehat{c} \in C$.

By restricting this representation to $\mathcal{D}$, from History Independence of Risk Preferences, without loss of generality we can set $U_{s^{t-1}}, u_{s^{t-1}}$ such that

$$U_{s^{t-1}}(d) = U(d), \quad u_{s^{t-1}}(c,d) = u(c,d)$$

holds for some common $U : \mathcal{D} \to \mathbb{R}$ and $u : C \times \mathcal{D} \to \mathbb{R}$ for all t and s^{t-1}.

Then, because

$$u(c,d) = W_{s^{t-1}}(c, u(\widehat{c}, d))$$

holds for all (c, d), the function W_s^{t-1} is independent of s^{t-1} and t. Hence we can take some $W : C \times range\ u(\widehat{c}, \cdot) \to range\ U$ such that

$$U_{s^{t-1}}(p) = \int_{C \times \mathcal{H}} W(c, u_{s^{t-1}}(\widehat{c}, h))dp(c, h)$$

holds for all t and s^{t-1}.

Define a preference relation $\succsim_{s^{t-1},\widehat{c}}$ over $\mathcal{H}$ by

$$h \succsim_{s^{t-1},\widehat{c}} h' \iff \delta(\widehat{c}, h) \succsim_{s^{t-1}} \delta(\widehat{c}, h').$$

Then it satisfies the Gilboa-Schmeidler set of axioms over the subdomain $\mathcal{H}_{+1} \subset \mathcal{H}$, and it is represented in the form

$$V_{s^{t-1}}(h_{+1}) = \min_{\mu \in M_{s^{t-1}}} \sum_{s \in S} v_{s^{t-1}}(h_{+1}(s))\mu(s),$$

where $V_{s^{t-1}} : \mathcal{H}_{+1} \to \mathbb{R}$ and $v_{s^{t-1}} : \mathcal{D} \to \mathbb{R}$ are unique up to positive affine transformations.

From History Independence of Risk Preference and Dynamic Consistency, $v_{s^{t-1}}$ satisfies Stationarity. Hence

$$\begin{aligned} v_{s^{t-1}}(d) \geq v_{s^{t-1}}(d') &\iff \delta(\widehat{c}, d) \succsim_{s^{t-1}} \delta(\widehat{c}, d') \\ &\iff \delta(\widehat{c}, d) \succsim \delta(\widehat{c}, d') \\ &\iff d \succsim d' \\ &\iff U(d) \geq U(d') \end{aligned}$$

for the common mixture-linear function U, and without loss of generality we can set $v_{s^{t-1}} = U$ so that

$$V_{s^{t-1}}(h_{+1}) = \min_{\mu \in M_{s^{t-1}}} \sum_{s \in S} U(h_{+1}(s))\mu(s)$$

holds.

Now extend $V_{s^{t-1}}$ to the entire $\mathcal{H}$ by

$$V_{s^{t-1}}(h) = V_{s^{t-1}}(h_{+1}),$$

where $h_{+1} \in \mathcal{H}_{+1}$ is such that

$$h(s) \sim_{s^{t-1},s} h_{+1}(s), \quad \forall s \in S, \quad \delta(\widehat{c}, h) \sim_{s^{t-1}} \delta(\widehat{c}, h_{+1}).$$

If you worry about the existence of such a one-step-ahead act, see Hayashi [57].

Given s^{t-1} and h, define an $|S|$-dimensional vector $U_{s^{t-1},h} \in \mathbb{R}^S$ by

$$U_{s^{t-1},h}(s) = U_{s^{t-1},s}(h(s)).$$

□

Lemma 8.10 *$V_{s^{t-1}}$ has the form*

$$V_{s^{t-1}}(h) = \Phi_{s^{t-1}}(U_{s^{t-1},h}),$$

where $\Phi_{s^{t-1}} : U(\mathcal{D})^S \to \mathbb{R}$ is strongly increasing.

Proof Let $U_{s^{t-1},h} = U_{s^{t-1},h'}$, then since $h(s) \sim_{s^{t-1},s} h'(s)$ holds for all $s \in S$, from Dynamic Consistency it follows that $\delta(\widehat{c}, h) \succsim_{s^{t-1}} \delta(\widehat{c}, h')$. Hence we have $V_{s^{t-1}}(h) = V_{s^{t-1}}(h')$ and the value of $V_{s^{t-1}}(h)$ depends only on the vector $U_{s^{t-1},h}$. Strong monotonicity follows similarly. □

Since $U_{s^{t-1},d} = U(d)\mathbf{1}$ and $V_{s^{t-1}}(d) = U(d)$, for all $a \in U(\mathcal{D})$,

$$\Phi_{s^{t-1}}(a\mathbf{1}) = a.$$

Lemma 8.11 *$\Phi_{s^{t-1}}$ satisfies the following properties:*

1.

$$\Phi_{s^{t-1}}(\lambda v + (1-\lambda)a\mathbf{1}) = \lambda\Phi_{s^{t-1}}(v) + (1-\lambda)a.$$

2.

$$\Phi_{s^{t-1}}(v) = \Phi_{s^{t-1}}(v') \implies \Phi_{s^{t-1}}(\lambda v + (1-\lambda)v') \geq \Phi_{s^{t-1}}(v).$$

Proof Property 1 follows from

$$\begin{aligned}
&\Phi_{s^{t-1}}(\lambda U_{s^{t-1},h_{+1}} + (1-\lambda)U_{s^{t-1},d}) \\
&\quad \text{from mixture independence of static risk applied at each state} \\
&= \Phi_{s^{t-1}}(\lambda U_{s^{t-1},\lambda h_{+1}+(1-\lambda d)}) \\
&= V_{s^{t-1}}(\lambda h_{+1} + (1-\lambda)d) \\
&\quad \text{from C-mixture linearity of} V_{s^{t-1}} \\
&= \lambda V_{s^{t-1}}(\lambda h_{+1}) + (1-\lambda)V_{s^{t-1}}(d) \\
&= \lambda\Phi_{s^{t-1}}(U_{s^{t-1},h_{+1}}) + (1-\lambda)\Phi_{s^{t-1}}(U_{s^{t-1},d}).
\end{aligned}$$

Property 2 follows, when $\Phi_{s^{t-1}}(U_{s^{t-1},h_{+1}}) = \Phi_{s^{t-1}}(U_{s^{t-1},h'_{+1}})$, from

$$\Phi_{s^{t-1}}(\lambda U_{s^{t-1},h_{+1}} + (1-\lambda)U_{s^{t-1},h'_{+1}})$$

from mixture independence of static risk applied at each state

$$= \Phi_{s^{t-1}}(\lambda U_{s^{t-1},\lambda h_{+1}+(1-\lambda h'_{+1})})$$

$$= V_{s^{t-1}}(\lambda h_{+1} + (1-\lambda)h'_{+1})$$

from super-additivity of $V_{s^{t-1}}$

$$\geq \lambda V_{s^{t-1}}(\lambda h_{+1}) + (1-\lambda)V_{s^{t-1}}(h'_{+1})$$

$$= \lambda \Phi_{s^{t-1}}(U_{s^{t-1},h_{+1}}) + (1-\lambda)\Phi_{s^{t-1}}(U_{s^{t-1},h'_{+1}}).$$

□

Without loss of generality, we can take $U(\underline{d}) < U(d^*) = 0 < U(\overline{d})$. Then we can extend $\Phi_{s^{t-1}}$ to the entire $\mathbb{R}^S$ so that the relation

$$\Phi_{s^{t-1}}(\lambda U_{s^{t-1},h_{+1}}) = \lambda \Phi_{s^{t-1}}(U_{s^{t-1},h_{+1}})$$

is maintained and

$$\mathbb{R}^S = \{\lambda U_{s^{t-1},h_{+1}} : h_{+1} \in \mathcal{H}_{+1}, \lambda \geq 0\}$$

is obtained.

Since the function $\Phi_{s^{t-1}}$ satisfies the Gilboa-Schmeidler type argument on $\mathbb{R}^S$, we obtain:

Lemma 8.12 *The function $\Phi_{s^{t-1}}$ has the form*

$$\Phi_{s^{t-1}}(U_{s^{t-1},h}) = \min_{\mu \in Q_{s^{t-1}}} \sum_{s \in S} U_{s^{t-1},h}(s)\mu(s).$$

On the subdomain $\mathcal{H}_{+1} \subset \mathcal{H}$ it holds $U_{s^{t-1},h_{+1}}(s) = U(h_{+1}(s))$ for all $s \in S$. Hence

$$V_{s^{t-1}}(h_{+1}) = \min_{\mu \in M_{s^{t-1}}} \sum_{s \in S} U(h_{+1}(s))\mu(s) = \min_{\mu \in Q_{s^{t-1}}} \sum_{s \in S} U(h_{+1}(s))\mu(s)$$

and uniqueness of $M_{s^{t-1}}$ implies $M_{s^{t-1}} = Q_{s^{t-1}}$. Hence we obtain

$$V_{s^{t-1}}(h) = \min_{\mu \in M_{s^{t-1}}} \sum_{s \in S} U_{s^{t-1},h}(s)\mu(s).$$

Since $u_{s^{t-1}}(\widehat{c}, \cdot)$ and $V_{s^{t-1}}(\cdot)$ represent the same ranking $\succsim_{s^{t-1}, \widehat{c}}$ over $\mathcal{H}$, there is a monotone transformation $\varphi_{s^{t-1}}$ such that

$$u_{s^{t-1}}(\widehat{c}, \cdot) = \varphi_{s^{t-1}}(V_{s^{t-1}}(\cdot))$$

holds.

By restricting this to $\mathcal{D}$, we see that the function $\varphi_{s^{t-1}}$ can be taken as a history-independent one, φ. Thus we have

$$\begin{aligned} U_{s^{t-1}}(p) &= \int_{C\times\mathcal{H}} W(c, u_{s^{t-1}}(\widehat{c}, h))dp(c, h) \\ &= \int_{C\times\mathcal{H}} W(c, \varphi(V_{s^{t-1}}(h)))dp(c, h) \end{aligned}$$

Finally by redefining W, we obtain

$$\begin{aligned} U_{s^{t-1}}(p) &= \int_{C\times\mathcal{H}} W(c, V_{s^{t-1}}(h))dp(c, h) \\ &= \int_{C\times\mathcal{H}} W\left(c, \min_{\mu\in M_{s^{t-1}}} \sum_{s\in S} U_{s^{t-1},s}(h(s))\mu(s)\right) dp(c, h). \end{aligned}$$

Uniqueness of $M_{s^{t-1}}$ has already been obtained. Uniqueness of $\{U_{s^{t-1}}\}$ up to positive affine transformations and uniqueness of W as stated in the theorem follows from the previous result obtained in the subdomain $\mathcal{D}$.

This completes the proof of the representation theorem. Now consider adding the axiom of timing indifference for risks.

Axiom 8.22 (Timing Indifference for Risks) For all $c \in C$, $d, d' \in \mathcal{D}$ and $\lambda \in [0, 1]$,

$$\lambda\delta(c, d) + (1-\lambda)\delta(c, d') \sim_{s^{t-1}} \delta(c, \lambda d + (1-\lambda)d')$$

Then we obtain that the aggregator function $W(\cdot, \cdot)$ is linear in the second argument.

Corollary 8.1 *Under the conditions as in the above theorem, the preference process satisfies Timing Indifference for Risks if and only is the aggregator function W has the form*

$$W(c, V) = u(c) + \beta(c)V.$$

Finally, let us confirm the three-way separation of elasticity of intertemporal substitution, risk aversion and ambiguity aversion.

Definition 8.2 Say that $\{\succsim^a_{s^{t-1}}\}$ is more risk-averse than $\{\succsim^b_{s^{t-1}}\}$ if for all s^{t-1} and $l \in \Delta(C^\infty)$, $y \in C^\infty$,

$$l \succsim^a_{s^{t-1}} y \implies l \succsim^b_{s^{t-1}} y$$

and

$$l \succ^a_{s^{t-1}} y \implies l \succ^b_{s^{t-1}} y.$$

Definition 8.3 Say that $\{\succsim^a_{s^{t-1}}\}$ is more ambiguity averse than $\{\succsim^b_{s^{t-1}}\}$ if for all s^{t-1} and $h_{+1} \in \mathcal{H}_{+1}$, $d \in \mathcal{D}$,

$$h_{+1} \succsim^a_{s^{t-1}} d \implies h_{+1} \succsim^b_{s^{t-1}} d$$

and

$$h_{+1} \succ^a_{s^{t-1}} d \implies h_{+1} \succ^b_{s^{t-1}} d$$

By applying the previous argument on separation of ambiguity aversion from risk aversion, and on separation of risk aversion from elasticity of intertemporal substitution, we obtain the result below.

Theorem 8.12 *Suppose* $(\{U^a_{s^{t-1}}, M^a_{s^{t-1}}\}, W^a)$ *represents* $\{\succsim^a_{s^{t-1}}\}$, *and* $(\{U^b_{s^{t-1}}, M^b_{s^{t-1}}\}, W^b)$ *represents* $\{\succsim^b_{s^{t-1}}\}$.

Then, $\{\succsim^a_{s^{t-1}}\}$ *is more ambiguity averse than* $\{\succsim^b_{s^{t-1}}\}$ *if and only if there exist* $A > 0$, $B \in \mathbb{R}$ *such that*

$$U^a_{s^{t-1}} = AU^b_{s^{t-1}} + B,$$
$$M^a_{s^{t-1}} \supset M^b_{s^{t-1}}$$

hold for all s^{t-1}, *and*

$$W^a(c, \widehat{V}) = AW^b\left(c, \frac{\widehat{V} - B}{A}\right) + B$$

for all $\widehat{V} \in \widehat{U}(D)$.

Chapter 9
Choice of Opportunity

9.1 "Rational" Choice of Opportunity

Let X be the set of final alternatives. Let $\mathcal{B}$ be the family of non-empty subsets of X, and let us call its element an **opportunity set**.

Choosing an opportunity set $A \in \mathcal{B}$ means to limit final alternatives to set A and leaves an opportunity that the final choice is made from A.

Preference $\succsim$ is defined over $\mathcal{B}$. Suppose $A \succsim B$ holds for $A, B \in \mathcal{B}$, then it means that the decision maker weakly prefers opportunity set A over opportunity set B.

"What's the point of thinking about this? Isn't it sufficient to consider preference just over final alternatives?" That is true, in the standard "rational" choice framework, in which direct utility maximization induces an indirect utility function. There, preference just over final alternatives is enough. Below is the formalization of such a standard argument.

Definition 9.1 Preference $\succsim$ is said to be **strategically rational** if

$$A \succsim B \implies A \cup B \sim A$$

holds for all $A, B \in \mathcal{B}$.

The above condition says that in "rational" choice, if one does not prefer an opportunity set (B here) it must be because it contains only inferior alternatives, and adding such set of alternatives to the better one (A here) should not benefit or harm either, since anything in B will never be chosen from $A \cup B$.

Under strategic rationality, we can show that preference over opportunity sets reduces to a preference ranking defined over final alternatives.

T. Hayashi, *Decision Theory*, Monographs in Mathematical Economics 9,
https://doi.org/10.1007/978-981-95-2200-2_9

Proposition 9.1 *$\succsim$ is strategically rational if and only if there is a preference relation $\succsim^*$ defined over X such that*

$$A \succsim B \iff t_{\succsim^*}(A) \succsim^* t_{\succsim^*}(B)$$

holds for all $A, B \in \mathcal{B}$, where $t_{\succsim^*}(A)$ is the maximal element for $\succsim^*$ in A, which is an arbitrary selection if there are several such ones.

Proof

$\Longleftarrow$ is obvious
$\Longrightarrow$: Define $\succsim^*$ by

$$x \succsim^* y \iff \{x\} \succsim \{y\}$$

for every $x, y \in X$. By repeated application of strategic rationality we obtain $\{t_{\succsim^*}(A)\} \sim A$, which delivers the conclusion.

□

9.2 Theory of Temptation and Self-control

9.2.1 Preference for Commitment

Let us consider the possibility that preference over opportunity sets does not reduce to a preference ranking over final alternatives.

To begin with, consider two final alternatives a and b. Then there are three possible opportunity sets, which are

$$\{a\}, \quad \{b\}, \quad \{a, b\}.$$

Here $\{a\}$ refers to an opportunity set that contains only a, and choosing such a set restricts your subsequent choices to only a. Similarly for $\{b\}$. On the other hand, choosing opportunity set $\{a, b\}$ leaves room to choose from a and b afterward.

Suppose for example that $\{a\} \succ \{b\}$. Then, as far as we have to make a perfect commitment, b is seen as an inferior alternative. Under strategic rationality, this immediately implies

$$\{a\} \sim \{a, b\}.$$

However, if the decision maker worries about a possibility that he himself mistakenly chooses b afterward when he leaves room to choose it, he would dislike to leave such an opportunity and we will have

$$\{a\} \succ \{a, b\}.$$

How about the relation between $\{a, b\}$ and $\{b\}$? If the decision maker is afraid that he will necessarily choose b when it is available, that is, he necessarily yields temptation, the ranking will be

$$\{a, b\} \sim \{b\}.$$

On the other hand, if he estimates that he will not necessarily yield to temptation the ranking will be

$$\{a, b\} \succ \{b\}.$$

In any case, in order to allow above possibilities, the ranking will be at least

$$\{a\} \succsim \{a, b\} \succsim \{b\}.$$

Note that there is *implicitly* a two-period model here, in which the decision maker chooses an opportunity set in Period 1 and he makes the final choice from it in Period 2. By "implicitly," I mean that the model only gives a ranking over opportunity sets and there is not an explicit model of choice from a given opportunity set in Period 2.

9.2.2 Model, Axioms and the Representation Theorem

Here we follow an axiomatic model due to Gul and Pesendorfer [51]. Let X be the set of final outcomes, which is assumed to be finite for simplicity. Let $\Delta(X)$ be the set of lotteries over X, which is a compact convex subset of an Euclidian space. Thinking of lotteries is only for making the argument operational, and there is not much to see conceptually.

Let $\mathcal{Z} = \mathcal{K}(\Delta(X))$ be the set of compact subsets of $\Delta(X)$. On $\mathcal{Z}$, we can define the Hausdorff metric

$$d(A, B) = \max\{\sup_{a \in A} \inf_{b \in B} d(a, b), \sup_{b \in B} \inf_{a \in A} d(a, b)\}$$

for every two sets $A, B \in \mathcal{Z}$. Also, the mixture operation over $\mathcal{Z}$ is defined by

$$\lambda A + (1 - \lambda)B = \{\lambda a + (1 - \lambda)b : a \in A,\ b \in B\}$$

for $A, B \in \mathcal{Z}$ and $\lambda \in [0, 1]$.

Preference $\succsim$ is defined over $\mathcal{Z}$, and we impose the following axioms.

Axiom 9.1 (Weak Order) $\succsim$ is complete and transitive.

Axiom 9.2 (Continuity) $\succsim$ is continuous with respect to the Hausdorff metric.

Axiom 9.3 (Independence) For all $A, B, C \in \mathcal{Z}$ and $\lambda \in (0, 1)$,

$$A \succsim B \iff \lambda A + (1 - \lambda)C \succsim \lambda B + (1 - \lambda)C.$$

The key axiom is betweenness.

Axiom 9.4 (Betweenness) For all $A, B \in \mathcal{Z}$,

$$A \succsim B \implies A \succsim A \cup B \succsim B.$$

This axiom allows the previous example

$$\{a\} \succsim \{a, b\} \succsim \{b\}$$

for a, b with $\{a\} \succsim \{b\}$.

Here is GP's main theorem.

Theorem 9.1 *$\succsim$ satisfies Continuity, Transitivity, Continuity, Independence and Betweenness if and only if there exist a mixture-linear and continuous function $U : \mathcal{Z} \to \mathbb{R}$ and mixture-linear continuous functions $u, v : \Delta(X) \to \mathbb{R}$ such that U represents $\succsim$ in the form*

$$U(A) = \max_{x \in A} \{u(x) + v(x)\} - \max_{y \in A} v(y),$$

where $A \in \mathcal{Z}$.

9.2.3 Key Arguments of the Proof

Necessity of the Axioms We prove necessity of Betweenness, that is, the only non-obvious one.

Proof Suppose $U(A) \geq U(B)$ and let

$$u(x_A) + v(x_A) = \max_{x \in A} \{u(x) + v(x)\}, \quad v(y_A) = \max_{y \in A} v(y),$$

$$u(x_B) + v(x_B) = \max_{x \in B} \{u(x) + v(x)\}, \quad v(y_B) = \max_{y \in A} v(y).$$

Then we have

$$\begin{aligned} U(A \cup B) &= \max_{x \in A \cup B} \{u(x) + v(x)\} - \max_{y \in A \cup B} v(y) \\ &= \max\{u(x_A) + v(x_A), u(x_B) + v(x_B)\} - \max\{v(y_A), v(y_B)\}. \end{aligned}$$

Now there are four cases.

Case 1: If $u(x_A) + v(x_A) \geq u(x_B) + v(x_B)$ and $v(y_A) \geq v(y_B)$, then we obtain

$$\begin{aligned} U(A \cup B) &= u(x_A) + v(x_A) - v(y_A) \\ &= U(A) \\ &\geq U(B). \end{aligned}$$

Case 2: If $u(x_A) + v(x_A) \leq u(x_B) + v(x_B)$ and $v(y_A) \leq v(y_B)$, then we obtain

$$\begin{aligned} U(A \cup B) &= u(x_B) + v(x_B) - v(y_B) \\ &= U(B) \\ &\leq U(A). \end{aligned}$$

Case 3: If $u(x_A) + v(x_A) \geq u(x_B) + v(x_B)$ and $v(y_A) \leq v(y_B)$, then we obtain

$$\begin{aligned} U(A \cup B) &= u(x_A) + v(x_A) - v(y_B) \\ &\geq u(x_B) + v(x_B) - v(y_B) \\ &= U(B) \end{aligned}$$

and

$$\begin{aligned} U(A \cup B) &= u(x_A) + v(x_A) - v(y_B) \\ &\leq u(x_A) + v(x_A) - v(y_A) \\ &= U(A). \end{aligned}$$

Case 4: If $u(x_A) + v(x_A) \leq u(x_B) + v(x_B)$ and $v(y_A) \geq v(y_B)$, then we obtain

$$\begin{aligned} U(A \cup B) &= u(x_B) + v(x_B) - v(y_A) \\ &\geq u(x_A) + v(x_A) - v(y_A) \\ &= U(A) \end{aligned}$$

and

$$\begin{aligned} U(A \cup B) &= u(x_B) + v(x_B) - v(y_A) \\ &\leq u(x_B) + v(x_B) - v(y_B) \\ &= U(B), \end{aligned}$$

which imply

$$U(A \cup B) \geq U(A) \geq U(B) \geq U(A \cup B),$$

which reduces to

$$U(A \cup B) = U(A) = U(B).$$

□

Sufficiency of the Axioms By a similar argument to the proof of Lemma 4.4, there is a mixture-linear function $U : \mathcal{Z} \to \mathbb{R}$, and we fix it. Then the commitment utility function $u : \Delta(X) \to \mathbb{R}$ is simply defined by

$$u(x) = U(\{x\}).$$

Since U is mixture-linear, u is mixture-linear on $\Delta(X)$.

The lemma below states that evaluation of opportunity sets reduces to that of two-point sets under Betweenness.

Lemma 9.1 *Suppose $\succsim$ satisfies Betweenness and it is represented by a function $U : \mathcal{Z} \to R$. Then for any finite opportunity set A,*

$$U(A) = \max_{x \in A} \min_{y \in A} U(\{x, y\}) = \min_{y \in A} \max_{x \in A} U(\{x, y\}).$$

Moreover, if (x^*, y^*) solves $\max_{x \in A} \min_{y \in A} U(\{x, y\})$ then (y^*, x^*) solves $\min_{y \in A} \max_{x \in A} U(\{x, y\})$.

Proof Denote any selection of the solutions to $\max_{x \in A} \min_{y \in A} U(\{x, y\})$ by (x^*, y^*). Then by definition, $U(\{x^*, y\}) \geq U(\{x^*, y^*\})$ holds for all $y \in A$. Thus by repeatedly applying Betweenness we obtain

$$U(A) = U\left(\bigcup_{y \in A} \{x^*, y\}\right) \geq U(\{x^*, y^*\}).$$

Also, for every $x \in A$ there exists $y_x \in A$ such that $U(\{x, y_x\}) \leq U(\{x^*, y^*\})$. Otherwise, there is $\widehat{x} \in A$ such that *any* $y \in A$ yields $U(\{\widehat{x}, y\}) > U(\{x^*, y^*\})$, which implies $\max_{x\in A}\min_{y\in A} U(\{x, y\}) \geq \min_{y\in A} U(\{\widehat{x}, y\}) > U(\{x^*, y^*\})$, a contradiction to the definition of (x^*, y^*).

Again by repeatedly applying Betweenness we obtain

$$U(A) = U\left(\bigcup_{x\in A}\{x, y_x\}\right) \leq U(\{x^*, y^*\})$$

and combining this with the previous claim we obtain

$$U(A) = U(\{x^*, y^*\}).$$

Now denote any selection of the solutions to $\min_{y\in A}\max_{x\in A} U(\{x, y\})$ by $(\widetilde{y}, \widetilde{x})$. Then by an argument analogous to that above we obtain

$$U(A) = U(\{\widetilde{x}, \widetilde{y}\})$$

and hence $U(\{x^*, y^*\}) = U(\{\widetilde{x}, \widetilde{y}\})$. Thus (y^*, x^*) solves $\min_{y\in A}\max_{x\in A} U(\{x, y\})$. □

Since evaluation of any opportunity sets reduces to that of two-point sets, once we obtain a claim on the subdomain of two-point sets saying

there exist mixture-linear functions $u, v : \Delta(X) \to \mathbb{R}$ such that

$$U(\{a, b\}) = \max_{w\in\{a,b\}}\{u(w) + v(w)\} - \max_{z\in\{a,b\}} v(z)$$

holds for all $a, b \in \Delta(X)$,

we can complete the proof of the theorem as follows. First we consider that A is a finite set. Then, from Lemma 9.1 we obtain

$$\begin{aligned} U(A) &= \max_{a\in A}\min_{b\in A} U(\{a, b\}) \\ &= \max_{a\in A}\min_{b\in A}\left\{\max_{w\in\{a,b\}}\{u(w) + v(w)\} - \max_{z\in\{a,b\}} v(z)\right\} \\ &= \max_{a\in A}\min_{b\in A}\left\{\max_{w\in\{a,b\}}\{u(w) + v(w)\} + \min_{z\in\{a,b\}}\{-v(z)\}\right\} \\ &= \max_{w\in A}\{u(w) + v(w)\} - \max_{z\in A} v(z). \end{aligned}$$

Now suppose A is any compact set, then we can take a sequence of finite sets $\{A^n\}$ that converges to A in the Hausdorff metric. Hence by Continuity we can extend the above representation to the whole $\mathcal{Z}$.

How do we obtain v then? To see this, consider a situation

$$U(\{a\}) > U(\{a, b\}) > U(\{b\}).$$

With the hindsight of the above representation, this means

$$u(a) > \max\{u(a) + v(a), u(b) + v(b)\} - \max\{v(a), v(b)\} > u(b).$$

Now, if $u(a) + v(a) \leq u(b) + v(b)$, then the inequality on the right implies

$$u(b) + v(b) - \max\{v(a), v(b)\} > u(b),$$

but this leads to

$$v(b) > \max\{v(a), v(b)\},$$

which is a contradiction. Also, if $v(a) \geq v(b)$, then the inequality on the left implies

$$u(a) > \max\{u(a) + v(a), u(b) + v(b)\} - v(a),$$

but this leads to

$$u(a) + v(a) > \max\{u(a) + v(a), u(b) + v(b)\},$$

which is again a contradiction. Hence a uniquely maximizes $u + v$ and b uniquely maximizes v.

Thus, looking at utility change according to changing b helps us to identify v. In fact, v is going to be defined by

$$v(x) = \frac{1}{\delta}\left(U(\{a, b\}) - U(\{a, (1 - \delta)b + \delta x\})\right).$$

The purpose of the next section is to show that the above definition does not depend on the choice of a or b or δ, and constitutes the representation indeed. Since it is a technical part, you may skip it if you are satisfied with the above outline.

9.2.4 Rest of the Proof of the Theorem

Lemma 9.2 *Let U be a mixture-linear representation of $\succsim$ that satisfies Betweenness. Then,*

$$U(\{x\}) > U(\{x, y\}) > U(\{y\}), \;\; U(\{a\}) > U(\{a, b\}) > U(\{b\})$$

imply

$$U(\lambda\{x, y\} + (1 - \lambda)\{a, b\}) = U(\{\lambda x + (1 - \lambda)a, \lambda y + (1 - \lambda)b\}).$$

Proof Consider the four-point set

$$\lambda\{x, y\}+(1-\lambda)\{a, b\} = \{\lambda x+(1-\lambda)a, \lambda x+(1-\lambda)b, \lambda y+(1-\lambda)a, \lambda y+(1-\lambda)b\}$$

Then, from Lemma 9.1, some $(w^*, z^*) \in \lambda\{x, y\} + (1-\lambda)\{a, b\}$ solves

$$\begin{aligned} & U(\lambda\{x, y\} + (1-\lambda)\{a, b\}) \\ = & \max_{w\in\lambda\{x,y\}+(1-\lambda)\{a,b\}} \min_{z\in\lambda\{x,y\}+(1-\lambda)\{a,b\}} U(\{w, z\}) \\ = & \min_{z\in\lambda\{x,y\}+(1-\lambda)\{a,b\}} \max_{w\in\lambda\{x,y\}+(1-\lambda)\{a,b\}} U(\{w, z\}). \end{aligned}$$

Now it suffices to show

$$(w^*, z^*) = (\lambda x + (1-\lambda)a, \lambda y + (1-\lambda)b).$$

First, from mixture-linearity,

$$U(\lambda\{x\} + (1-\lambda)\{a, b\}) > U(\lambda\{x, y\} + (1-\lambda)\{a, b\}) > U(\lambda\{y\} + (1-\lambda)\{a, b\})$$

and

$$U(\lambda\{x, y\} + (1-\lambda)\{a\}) > U(\lambda\{x, y\} + (1-\lambda)\{a, b\}) > U(\lambda\{x, y\} + (1-\lambda)\{b\}).$$

Suppose $w^* = \lambda x+(1-\lambda)b$, then from the maxmin condition and the right-hand inequality of the second line above we obtain

$$\begin{aligned} U(\lambda\{x, y\} + (1-\lambda)\{a, b\}) &\le U(\lambda\{x, y\} + (1-\lambda)\{b\}) \\ &< U(\lambda\{x, y\} + (1-\lambda)\{a, b\}), \end{aligned}$$

which is a contradiction.

Likewise, suppose $w^* = \lambda y + (1-\lambda)b$, then again from the maxmin condition and the right-hand inequality of the second line above we obtain

$$\begin{aligned} U(\lambda\{x, y\} + (1-\lambda)\{a, b\}) &\le U(\lambda\{x, y\} + (1-\lambda)\{b\}) \\ &< U(\lambda\{x, y\} + (1-\lambda)\{a, b\}), \end{aligned}$$

which is a contradiction.

Finally, suppose $w^* = \lambda y + (1-\lambda)a$, then from the maxmin condition and the right-hand inequality of the first line above we obtain

$$\begin{aligned} U(\lambda\{x, y\} + (1-\lambda)\{a, b\}) &\le U(\lambda\{y\} + (1-\lambda)\{a, b\}) \\ &< U(\lambda\{x, y\} + (1-\lambda)\{a, b\}), \end{aligned}$$

which is a contradiction. Hence $w^* = \lambda x + (1-\lambda)a$.

Now, suppose $z^* = \lambda x + (1-\lambda)b$, then from the maxmin condition and the left-hand inequality of the first line above we obtain

$$\begin{aligned} U(\lambda\{x, y\} + (1-\lambda)\{a, b\}) &\geq U(\lambda\{x\} + (1-\lambda)\{a, b\}) \\ &> U(\lambda\{x, y\} + (1-\lambda)\{a, b\}), \end{aligned}$$

which is a contradiction.

Likewise, suppose $z^* = \lambda x + (1-\lambda)a$, then again from the maxmin condition and the left-hand inequality of the first line above we obtain

$$\begin{aligned} U(\lambda\{x, y\} + (1-\lambda)\{a, b\}) &\geq U(\lambda\{x\} + (1-\lambda)\{a, b\}) \\ &> U(\lambda\{x, y\} + (1-\lambda)\{a, b\}), \end{aligned}$$

which is a contradiction.

Finally, suppose $z^* = \lambda y + (1-\lambda)a$, then from the maxmin condition and the left-hand inequality of the second line we obtain

$$\begin{aligned} U(\lambda\{x, y\} + (1-\lambda)\{a, b\}) &\geq U(\lambda\{x, y\} + (1-\lambda)\{a\}) \\ &> U(\lambda\{x, y\} + (1-\lambda)\{a, b\}), \end{aligned}$$

which is a contradiction. Hence $z^* = \lambda y + (1-\lambda)b$. □

Now, given $a, b \in \Delta(X)$ and $\delta \in (0, 1)$, and define $v(\cdot; a, b, \delta) : \Delta(X) \to \mathbb{R}$ by

$$v(x; a, b, \delta) = \frac{U(\{a, b\}) - U(\{a, (1-\delta)b + \delta x\})}{\delta}.$$

Lemma 9.3 *Let U be a mixture-linear representation of $\succsim$ that satisfies Betweenness. Also, assume that*

$$U(\{a\}) > U(\{a, (1-\delta)b + \delta z\}) > U(\{(1-\delta)b + \delta z\})$$

holds for all $z \in \Delta(X)$. Then,

(i) For all $z \in \Delta(X)$ satisfying $U(\{a\}) > U(\{a, z\}) > U(\{z\})$,

$$v(z; a, b, \delta) = U(\{a, b\}) - U(\{a, z\})$$

holds for all $\delta \in (0, 1)$.

(ii) $v(a; a, b, \delta) = U(\{a, b\}) - U(\{a\})$

(iii) $v(\alpha z + (1-\alpha)z'; a, b, \delta) = \alpha v(z; a, b, \delta) + (1-\alpha)v(z'; a, b, \delta)$

(iv) For all $\delta' \in (0, \delta)$, $v(z; a, b, \delta') = v(z; a, b, \delta)$.

(v) *If* $U(\{x\}) > U(\{x, (1-\delta)y+\delta z\}) > U(\{(1-\delta)y+\delta z\})$ *holds for all* $z \in \Delta(X)$, *then* $v(z; a, b, \delta) = v(z; x, y, \delta) + v(y; a, b, \delta)$.

Proof

(i): Since the presumption of Lemma 9.2 is met, we have

$$(1-\delta)U(\{a, b\}) + \delta U(\{a, z\}) = U(\{a, (1-\delta)b + \delta z\}),$$

and this implies

$$\frac{U(\{a, b\}) - U(\{a, (1-\delta)b + \delta z\})}{\delta} = U(\{a, b\}) - U(\{a, z\}).$$

(ii): From mixture-linearity of U,

$$(1-\delta)U(\{a, b\}) + \delta U(\{a\}) = U(\{a, (1-\delta)b + \delta a\}),$$

hence we obtain

$$\frac{U(\{a, b\}) - U(\{a, (1-\delta)b + \delta a\})}{\delta} = U(\{a, b\}) - U(\{a\}).$$

(iii): It follows from Lemma 9.2 and mixture-linearity of U.
(iv): It suffices to see that the argument in (i) does not depend on the choice of δ.
(v): What is required to show is

$$\begin{aligned} &\frac{U(\{a, b\}) - U(\{a, (1-\delta)b + \delta z\})}{\delta} \\ =\ &\frac{U(\{x, y\}) - U(\{x, (1-\delta)y + \delta z\})}{\delta} \\ &+ \frac{U(\{a, b\}) - U(\{a, (1-\delta)b + \delta y\})}{\delta} \end{aligned}$$

and this is equivalent to

$$\begin{aligned} &U(\{x, (1-\delta)y + \delta z\}) + U(\{a, (1-\delta)b + \delta y\}) \\ =\ &U(\{x, y\}) + U(\{u, (1-\delta)b + \delta z\}). \end{aligned}$$

From mixture-linearity this is equivalent to

$$\begin{aligned} &U\left(\frac{1}{2}\{x, (1-\delta)y + \delta z\} + \frac{1}{2}\{a, (1-\delta)b + \delta y\}\right) \\ =\ &U\left(\frac{1}{2}\{x, y\} + \frac{1}{2}\{a, (1-\delta)b + \delta z\}\right). \end{aligned}$$

From Lemma 9.2, both sides of the above are equal to

$$U\left(\left\{\frac{x+a}{2}, \frac{(1-\delta)b+\delta z+y}{2}\right\}\right).$$

□

Lemma 9.4 *Let U be a mixture-linear representation of $\succsim$ that satisfies Betweenness. Consider $a, y \in \Delta(X)$ such that $U(\{a\}) \geq U(\{a, y\}) \geq U(\{y\})$ holds.*

Suppose $b \in \Delta(X)$ and $\delta \in (0, 1)$ are such that

$$U(\{a\}) > U(\{a, (1-\delta)b+\delta z\}) > U(\{(1-\delta)b+\delta z\})$$

holds for all $z \in \Delta(X)$.

Then,

$$U(\{a, y\}) = \max_{w\in\{a,y\}} \{u(w) + v(w; a, b, \delta)\} - \max_{z\in\{a,y\}} v(z; a, b, \delta).$$

Proof **Case 1**: Suppose $U(\{a\}) > U(\{a, y\}) > U(\{y\})$. Then from Lemma 9.3 (i) and (ii), we have

$$v(y; a, b, \delta) = U(\{a, b\}) - U(\{a, y\}) > U(\{a, b\}) - U(\{a\}) = v(a; a, b, \delta)$$

and $\max_{z\in\{a,y\}} v(z; a, b, \delta) = v(y; a, b, \delta)$ follows. Also, from (i) and (ii) we have

$$\begin{aligned}
&u(a) + v(a; a, b, \delta) - v(y; a, b, \delta)\\
&= U(\{a\}) + U(\{a, b\}) - U(\{a\}) - U(\{a, b\}) + U(\{a, y\})\\
&= U(\{a, y\})\\
&> U(\{y\})\\
&= u(y)\\
&= u(y) + v(y; a, b, \delta) - v(y; a, b, \delta)
\end{aligned}$$

and $\max_{w\in\{a,y\}}\{u(w) + v(w; a, b, \delta)\} = u(a) + v(a; a, b, \delta)$ follows. Thus we obtain

$$\begin{aligned}
&U(\{a, y\})\\
&= u(a) + v(a; a, b, \delta) - v(y; a, b, \delta)\\
&= \max_{w\in\{a,y\}} \{u(w) + v(w; a, b, \delta)\} - \max_{z\in\{a,y\}} v(z; a, b, \delta).
\end{aligned}$$

Case 2: Suppose $U(\{a\}) = U(\{a, y\}) > U(\{y\})$. Then the conclusion follows if we can show $v(a; a, b, \delta) \geq v(y; a, b, \delta)$, since

$$\begin{aligned}
&\max_{w\in\{a,y\}} \{u(w) + v(w; a, b, \delta)\} - \max_{z\in\{a,y\}} v(z; a, b, \delta) \\
&= u(a) + v(a; a, b, \delta) - v(a; a, b, \delta) \\
&= u(a) \\
&= U(\{a\}) \\
&= U(\{a, y\}).
\end{aligned}$$

We will show $v(a; a, b, \delta) \geq v(y; a, b, \delta)$. By definition, this means

$$\frac{U(\{a, b\}) - U(\{a, (1-\delta)b + \delta a\})}{\delta} \geq \frac{U(\{a, b\}) - U(\{a, (1-\delta)b + \delta y\})}{\delta}.$$

From mixture-linearity of U and the current assumption $U(\{a\}) = U(\{a, y\})$, we have

$$\begin{aligned}
U(\{a, (1-\delta)b + \delta a\}) &= (1-\delta)U(\{a, b\}) + \delta U(\{a\}) \\
&= (1-\delta)U(\{a, b\}) + \delta U(\{a, y\}) \\
&= U((1-\delta)\{a, b\} + \delta\{a, y\})
\end{aligned}$$

Therefore what we have to show is

$$U((1-\delta)\{a, b\} + \delta\{a, y\}) \leq U(\{a, (1-\delta)b + \delta y\}).$$

Once we can show

$$U((1-\delta)\{a, b\} + \delta\{a, y\}) = \min_{w\in(1-\delta)\{a,b\}+\delta\{a,y\}} U(\{a, w\}),$$

the above condition is implied because $(1-\delta)b + \delta y \in (1-\delta)\{a, b\} + \delta\{a, y\}$.

From Lemma 9.1, there exists $z \in (1-\delta)\{a, b\} + \delta\{a, y\}$ such that

$$U((1-\delta)\{a, b\} + \delta\{a, y\}) = \min_{w\in(1-\delta)\{a,b\}+\delta\{a,y\}} U(\{z, w\}).$$

Now we show it has to be that $z = a$.

Suppose $z \neq a$, then there are three cases.

Suppose $z = (1-\delta)a + \delta y$, then by taking $w = (1-\delta)b + \delta y$, from mixture-linearity and the presumption $U(\{a, y\}) > U(\{y\})$, we have

$$\begin{aligned}
U((1-\delta)\{a, b\} + \delta\{a, y\}) &= \min_{w \in (1-\delta)\{a,b\}+\delta\{a,y\}} U(\{z, w\}) \\
&\leq U(\{(1-\delta)a + \delta y, (1-\delta)b + \delta y\}) \\
&= (1-\delta)U(\{a, b\}) + \delta U(\{y\}) \\
&< (1-\delta)U(\{a, b\}) + \delta U(\{a, y\}),
\end{aligned}$$

which is a contradiction to mixture-linearity.

Next, suppose $z = (1-\delta)b + \delta a$, then by taking $w = (1-\delta)b + \delta y$, from mixture-linearity and $U(\{a, b\}) > U(\{b\})$ that follows from the assumption, we have

$$\begin{aligned}
U((1-\delta)\{a, b\} + \delta\{a, y\}) &= \min_{w \in (1-\delta)\{a,b\}+\delta\{a,y\}} U(\{z, w\}) \\
&\leq U(\{(1-\delta)b + \delta a, (1-\delta)b + \delta y\}) \\
&= (1-\delta)U(\{b\}) + \delta U(\{a, y\}) \\
&< (1-\delta)U(\{a, b\}) + \delta U(\{a, y\}),
\end{aligned}$$

which is a contradiction to mixture-linearity.

Finally, suppose $z = (1-\delta)b + \delta y$, then by taking $w = (1-\delta)b + \delta y$, from mixture-linearity and $U(\{a, b\}) > U(\{b\})$ and $U(\{a, y\}) > U(\{y\})$ that follow from the assumption, we have

$$\begin{aligned}
U((1-\delta)\{a, b\} + \delta\{a, y\}) &= \min_{w \in (1-\delta)\{a,b\}+\delta\{a,y\}} U(\{z, w\}) \\
&\leq U(\{(1-\delta)b + \delta y\}) \\
&= (1-\delta)U(\{b\}) + \delta U(\{y\}) \\
&< (1-\delta)U(\{a, b\}) + \delta U(\{a, y\}),
\end{aligned}$$

which is a contradiction to mixture-linearity.

Case 3: Suppose $U(\{a\}) > U(\{a, y\}) = U(\{y\})$. Then, since $u(a) = U(\{a\}) > U(\{y\}) = u(y)$ follows, once we show $v(y; a, b, \delta) \geq u(a) + v(a; a, b, \delta) - u(y)$ it follows that $v(y; a, b, \delta) > v(a; a, b, \delta)$, and we will have the conclusion because

$$\begin{aligned}
&\max_{w \in \{a,y\}} \{u(w) + v(w; a, b, \delta)\} - \max_{z \in \{a,y\}} v(z; a, b, \delta) \\
&= u(y) + v(y; a, b, \delta) - v(y; a, b, \delta) \\
&= u(y)
\end{aligned}$$

$$= U(\{y\})$$
$$= U(\{a, y\}).$$

From Lemma 9.3 (ii) and the assumption $U(\{a, y\}) = U(\{y\})$, we obtain

$$\begin{aligned} u(a) + v(a; a, b, \delta) - u(y) &= U(\{a\}) + U(\{a, b\}) - U(\{a\}) - U(\{y\}) \\ &= U(\{a, b\}) - U(\{y\}) = U(\{a, b\}) - U(\{a, y\}). \end{aligned}$$

Hence, by going back to the definition of $v(y; a, b, \delta)$, then the condition we are to show takes the form

$$\frac{U(\{a, b\}) - U(\{a, (1-\delta)b + \delta y\})}{\delta} \geq U(\{a, b\}) - U(\{a, y\}),$$

which is rewritten as

$$(1-\delta)U(\{a, b\}) + \delta U(\{a, y\}) \geq U(\{a, (1-\delta)b + \delta y\}).$$

Now, from mixture-linearity and Lemma 9.1 it follows that

$$\begin{aligned} &(1-\delta)U(\{a, b\}) + \delta U(\{a, y\}) \\ &= U((1-\delta)\{a, b\} + \delta\{a, y\}) \\ &= \max_{z \in (1-\delta)\{a,b\})+\delta\{a,y\}} \min_{w \in (1-\delta)\{a,b\})+\delta\{a,y\}} U(\{z, w\}) \\ &\geq \min_{w \in (1-\delta)\{a,b\})+\delta\{a,y\}} U(\{a, w\}). \end{aligned}$$

Because

$$\begin{aligned} U((1-\delta)\{a, b\} + \delta\{a, y\}) &= (1-\delta)U(\{a, b\}) + \delta U(\{a, y\}) \\ &< (1-\delta)U(\{a\}) + \delta U(\{a, y\}) \\ &= U(\{a, (1-\delta)a + \delta y\}) \end{aligned}$$

and

$$\begin{aligned} U((1-\delta)\{a, b\} + \delta\{a, y\}) &= (1-\delta)U(\{a, b\}) + \delta U(\{a, y\}) \\ &< (1-\delta)U(\{a, b\}) + \delta U(\{a\}) \\ &= U(\{a, (1-\delta)b + \delta a\}) \end{aligned}$$

and

$$\begin{aligned} U((1-\delta)\{a,b\}+\delta\{a,y\}) &= (1-\delta)U(\{a,b\})+\delta U(\{a,y\}) \\ &< (1-\delta)U(\{a\})+\delta U(\{a\}) \\ &= U(\{a\}) \end{aligned}$$

we have

$$\min_{w\in(1-\delta)\{a,b\})+\delta\{a,y\}} U(\{a,w\}) = U(\{a,(1-\delta)b+\delta y\}).$$

Thus we obtain

$$(1-\delta)U(\{a,b\})+\delta U(\{a,y\}) \geq U(\{a,(1-\delta)b+\delta y\}).$$

Case 4: Suppose $U(\{a\}) = U(\{a,y\}) = U(\{y\})$. Then we can take $u(a) = u(y) = u^*$ for some constant u^*, so that we obtain

$$U(\{a,y\}) = u^* = \max_{w\in\{a,y\}}\{u^* + v(w;a,b,\delta)\} - \max_{z\in\{a,y\}} v(z;a,b,\delta).$$

□

Lemma 9.5 *There exists a pair of mixture-linear functions $u, v : \Delta(X) \to \mathbb{R}$ such that*

$$U(\{a,b\}) = \max_{w\in\{a,b\}}\{u(w)+v(w)\} - \max_{z\in\{a,b\}} v(z)$$

holds for all $a, b \in \Delta(X)$.

Proof If $U(\{a\}) = U(\{b\})$ for all $a, b \in \Delta(X)$, then by Betweenness it follows that $U(\{a,b\}) = U(\{a\}) = U(\{b\})$, so that the conclusion is obvious. Hence we assume there are $a, b \in \Delta(X)$ such that $U(\{a\}) > U(\{b\})$. Below, we consider four cases.

Case 1: Suppose there exist $x^*, y^* \in \Delta(X)$ such that

$$U(\{x^*\}) > U(\{x^*, y^*\}) > U(\{y^*\})$$

holds. Then, by continuity and compactness of $\Delta(X)$, there exists $\delta > 0$ such that

$$U(\{x^*\}) > U(\{x^*, (1-\delta)y^* + \delta z\}) > U(\{(1-\delta)y^* + \delta z\})$$

holds for all $z \in \Delta(X)$.

Now define

$$u(z) = U(\{z\}),\quad v(z) = v(z; x^*, y^*, \delta).$$

Consider any two points in the interior of $\Delta(X)$, denoted by $a, b \in int\Delta(X)$. Without loss of generality, assume $u(a) \geq u(b)$. Since a is in the interior of $\Delta(X)$, there are some $a' \in \Delta(X)$ and $\alpha \in (0, 1)$ such that $\alpha a' + (1-\alpha)x^* = a$. Then, by letting $c = \alpha a' + (1-\alpha)y^*$ we have

$$\begin{aligned} U(\{a\}) &= U(\{\alpha a' + (1-\alpha)x^*\}) \\ &= \alpha U(\{a'\}) + (1-\alpha)U(\{x^*\}) \\ &> \alpha U(\{a'\}) + (1-\alpha)U(\{x^*, y^*\}) \\ &= U(\{\alpha a' + (1-\alpha)x^*, \alpha a' + (1-\alpha)y^*\}) \\ &= U(\{a, c\}) \end{aligned}$$

and

$$\begin{aligned} U(\{c\}) &= U(\{\alpha a' + (1-\alpha)y^*\}) \\ &= \alpha U(\{a'\}) + (1-\alpha)U(\{y^*\}) \\ &< \alpha U(\{a'\}) + (1-\alpha)U(\{x^*, y^*\}) \\ &= U(\{\alpha a' + (1-\alpha)x^*, \alpha a' + (1-\alpha)y^*\}) \\ &= U(\{a, c\}), \end{aligned}$$

which imply

$$U(\{a\}) > U(\{a, c\}) > U(\{c\}).$$

By Continuity, for sufficiently small δ'

$$U(\{a\}) > U(\{a, (1-\delta')c + \delta' z\}) > U(\{(1-\delta')c + \delta' z\})$$

holds for all $z \in \Delta(X)$.

Hence by Lemma 9.4 we obtain

$$U(\{a, b\}) = \max_{w \in \{a,b\}} \{u(w) + v(w; a, c, \delta')\} - \max_{z \in \{a,b\}} v(z; a, c, \delta').$$

Let $\delta'' = \min\{\delta, \delta'\}$, then from the previous lemma there is a constant K such that

$$\begin{aligned} v(\cdot; a, c, \delta') &= v(\cdot; a, c, \delta'') = v(\cdot; x^*, y^*, \delta'') + K = v(\cdot; x^*, y^*, \delta) + K \\ &= v(\cdot) + K \end{aligned}$$

holds. Thus we obtain

$$U(\{a,b\}) = \max_{w\in\{a,b\}} \{u(w)+v(w)\} - \max_{z\in\{a,b\}} v(z).$$

By continuity, we can extend the above to any $a, b \in \Delta(X)$ which are not necessarily in the interior.

Case 2: Suppose

$$U(\{x\}) > U(\{y\}) \implies U(\{x\}) = U(\{x,y\}) > U(\{y\})$$

holds for all $x, y \in \Delta(X)$. In this case we set $u(x) = U(\{x\})$ and $v(x) = 0$.

Case 3: Suppose

$$U(\{x\}) > U(\{y\}) \implies U(\{x\}) > U(\{x,y\}) = U(\{y\})$$

holds for all $x, y \in \Delta(X)$. In this case we set $u(x) = U(\{x\})$ and $v(x) = -U(\{x\})$.

Case 4: Suppose that for all $x, y \in \Delta(X)$ either

$$U(\{x\}) > U(\{y\}) \implies U(\{x\}) = U(\{x,y\}) > U(\{y\})$$

or

$$U(\{x\}) > U(\{y\}) \implies U(\{x\}) > U(\{x,y\}) = U(\{y\}).$$

Then, pick $x, y \in \Delta(X)$ with $U(\{x\}) = U(\{x,y\}) > U(\{y\})$, and pick $a, b \in \Delta(X)$ with $U(\{a\}) > U(\{a,b\}) = U(\{b\})$. Then, the function defined by

$$f(\alpha) = \frac{U(\{\alpha x + (1-\alpha)a, \alpha y + (1-\alpha)b\}) - (\alpha U(\{y\}) + (1-\alpha)U(\{b\}))}{\alpha U(\{x\}) + (1-\alpha)U(\{a\}) - (\alpha U(\{y\}) + (1-\alpha)U(\{b\}))}$$

satisfies $f(0) = 0$ and $f(1) = 1$. Hence by continuity there is $\alpha^* \in (0,1)$ such that $f(\alpha^*) = 0.5$. For such α^*,

$$\begin{aligned} U(\{\alpha^* x + (1-\alpha^*)a\}) &> U(\{\alpha^* x + (1-\alpha^*)a, \alpha^* y + (1-\alpha^*)b\}) \\ &> U(\{\alpha^* y + (1-\alpha^*)b\}) \end{aligned}$$

and it reduces to Case 1. □

9.2.5 Dynamic Extension

Here we briefly see how the model of self-control as above can be extended to dynamic environments, following Gul and Pesendorfer [52].

By following a way similar to what we did in Chap. 8, we can establish a recursive domain of intertemporal choice problems $\mathcal{Z}$, with a slight abuse of notation by using the same character as before, which satisfies the recursive homeomorphism

$$\mathcal{Z} \simeq \mathcal{K}(\Delta(C \times \mathcal{Z}))$$

and allows mixture operations just as in the static setting. When we have a choice problem $A \in \mathcal{Z}$, it is a (compact) set of lotteries over pairs of current consumption and choice problem in the next period.

Preference $\succsim$ is defined over $\mathcal{Z}$, and we continue to impose the following axioms.

Axiom 9.5 (Weak Order) $\succsim$ is complete and transitive.

Axiom 9.6 (Continuity) $\succsim$ is continuous with respect to the Hausdorff metric.

Axiom 9.7 (Independence) For all $A, B, C \in \mathcal{Z}$ and $\lambda \in (0, 1)$,

$$A \succsim B \iff \lambda A + (1 - \lambda)C \succsim \lambda B + (1 - \lambda)C.$$

Axiom 9.8 (Betweenness) For all $A, B \in \mathcal{Z}$,

$$A \succsim B \implies A \succsim A \cup B \succsim B.$$

By applying the preceding theorem, we see that there exist a mixture-linear and continuous function $U : \mathcal{Z} \to \mathbb{R}$ and mixture-linear continuous functions $u, v : \Delta(C \times \mathcal{Z}) \to \mathbb{R}$ such that U represents $\succsim$ in the form

$$U(A) = \max_{l \in A} \{u(l) + v(l)\} - \max_{m \in A} v(m),$$

where $A \in \mathcal{Z}$.

Now we add elements which are specific to the intertemporal setting. First is rather for simplicity, which states that the decision maker does not care for correlation between current consumption and facing a choice problem in the next period, when it comes to choice of a one-step-ahead lottery with commitment. Let $\delta[c, A]$ denote the lottery degenerate on $(c, A) \in C \times \mathcal{Z}$.

Axiom 9.9 (Separability) For all $c, d \in C$ and $A, B \in \mathcal{Z}$,

$$\left\{\frac{1}{2}\delta[c, A] + \frac{1}{2}\delta[d, B]\right\} \sim \left\{\frac{1}{2}\delta[c, B] + \frac{1}{2}\delta[d, A]\right\}.$$

By Separability, the commitment utility function u is decomposed into two mixture-linear and continuous functions

$$u(l) = u_1(l_1) + u_2(l_2),$$

where l_1 denotes the marginal of lottery $l \in \Delta(C \times \mathcal{Z})$ over C and l_2 denotes the marginal over $\mathcal{Z}$.

Stationarity states that as far as the decision maker can commit to the choice of a choice problem in the next period there is no preference reversal.

Axiom 9.10 (Stationarity) For all $c \in C$ and $A, B \in \mathcal{Z}$,

$$\{\delta[c, A]\} \succsim \{\delta[c, B]\}$$

if and only if

$$A \succsim B.$$

By Stationarity, both u_2 restricted to degenerate lotteries over $\mathcal{Z}$ and U represent the same ranking over $\mathcal{Z}$. Hence we have ordinal equivalence

$$u_2(\delta[A]) = \phi(U(A)).$$

Axiom 9.11 (Timing Indifference) For all $c \in C, \alpha \in [0, 1]$ and $A, B \in \mathcal{Z}$,

$$\{\alpha\delta[c, A] + (1-\alpha)\delta[c, B]\} \sim \{\delta[c, \alpha A + (1-\alpha)B]\}.$$

By Timing Indifference, we have

$$\begin{aligned}
\phi(\alpha U(A) + (1-\alpha)U(B)) &= \phi(U(\alpha A + (1-\alpha)B)) \\
&= u_2(\delta[A + (1-\alpha)B]) \\
&= u_2(\delta[A] + (1-\alpha)\delta[B]) \\
&= \alpha u_2(\delta[A]) + (1-\alpha)u_2(\delta[B]) \\
&= \alpha\phi(U(A)) + (1-\alpha)\phi(U(B)),
\end{aligned}$$

where the axiom is applied between line 2 and line 3 above, which shows that ϕ is mixture-linear. Thus we have

$$u_2(\delta[A]) = \beta U(A) + \gamma,$$

with $\beta > 0$, here we ignore the constant γ without loss of generality. By the previous argument on continuity in Chap. 8, we have $\beta < 1$.

Thus the commitment utility function is written in the form

$$u(l) = u_1(l_1) + \beta \int_{\mathcal{Z}} U(A) dl_2(A).$$

Now rewrite u_1 as u with an abuse of notation.

The last axiom states that temptation is coming only from current consumption.

Axiom 9.12 (Temptation by Immediate Consumption) For all $l, m, n \in \Delta(C \times \mathcal{Z})$ with $m_1 = n_1$, if $\{l\} \succ \{l, m\} \succ \{m\}$ and $\{l\} \succ \{l, n\} \succ \{n\}$ then $\{l, m\} \sim \{l, n\}$.

By Temptation by Immediate Consumption, $v(l)$ depends only on l_1. Thus we write $v(l_1)$ without loss of generality.

As we take necessity of the axioms to be straightforward, now we have the theorem as below.

Theorem 9.2 *$\succsim$ satisfies Continuity, Transitivity, Continuity, Independence, Betweenness, Separability, Stationarity, Timing Indifference and Temptation by Immediate Consumption if and only if there exist a mixture-linear and continuous functions $U : \mathcal{Z} \to \mathbb{R}$ and mixture-linear and continuous functions $u, v : \Delta(C) \to \mathbb{R}$ and $\beta \in (0, 1)$ such that U represents $\succsim$ in the form*

$$U(A) = \max_{l \in A} \left\{ u(l_1) + v(c) + \beta \int_{\mathcal{Z}} U(A) dl_2(A) \right\} - \max_{m \in A} v(m_1),$$

where $A \in \mathcal{Z}$.

The last axiom, which excludes temptation by future consumptions, might look restrictive since the decision maker might be tempted by a life course itself instead. Noor [84] axiomatically characterized a class of dynamic self-control preferences in which the decision maker is tempted by an alternative life course and such temptation utility itself follows the stationary discounted utility. That is, Noor characterized the temptation representation

$$U(A) = \max_{l \in A} \left\{ u(l_1) + \beta \int_{\mathcal{Z}} U(A) dl_2(A) + V(l) \right\} - \max_{m \in A} V(m)$$

in which the temptation utility term satisfies the recursive formula

$$V(l) = \int_{C \times \mathcal{Z}} \{v(c) + \gamma V(B)\} dl(c, B)$$

with $\gamma \in (0, 1)$. Note, however, that this class is disjoint with the class of self-control preferences satisfying Temptation by Immediate Consumptions because $\gamma > 0$.

9.3 Preference for Flexibility and "Subjective State Space"*

In the previous section, we considered preference for narrowing an opportunity set for the sake of commitment. On the other hand, it is worth considering preference for leaving a larger opportunity, that is, ranking such as

$$A \succsim B \text{ and } A \cup B \succ A.$$

From the fact that the decision maker prefers to leave a larger opportunity, we can infer that he wants to keep an alternative available to choose, as there might be a possibility that choosing it is better, while he does not find the value to choose it with commitment. Thus we can infer existence of an ex-post state of the world that is subjectively perceived by the decision maker at which such alternative is preferred.

This is somehow tautological, since an ex-post state at which the decision maker prefers such an alternative is nothing but an ex-post preference which likes the alternative, and we won't able to physically identify such a state.

The idea dates back to Kreps [71], while here we follow Dekel et al. [24], which is more operational because it adopts the lottery device.

Let X be the set of final outcomes, which is assumed to be finite. Let $\Delta(X)$ be the set of lotteries over X, then it is a compact and convex subset of an Euclidian space. Again, thinking of lotteries is only for making the argument operational, and there is not much to see conceptually.

Let $\mathcal{Z}$ be the set of compact subsets of $\Delta(X)$, which is endowed with the Hausdorff metric. Also, let $\mathcal{K}$ be the set of compact and convex subsets of $\Delta(X)$. The preference relation $\succsim$ is defined over $\mathcal{Z}$.

Let us consider the following axioms.

Axiom 9.13 (Weak Order) $\succsim$ is complete and transitive.

Axiom 9.14 (Continuity) $\succsim$ is continuous with respect to the Hausdorff metric.

Axiom 9.15 (Mixture Independence) For all $A, B, C \in \mathcal{Z}$ and $\lambda \in (0, 1)$,

$$A \succsim B \iff \lambda A + (1 - \lambda)C \succsim \lambda B + (1 - \lambda)C.$$

As we follow the idea of preference for flexibility, it will be reasonable to impose monotonicity.

Axiom 9.16 (Monotonicity) For all $A, B \in \mathcal{Z}$,

$$A \supset B \implies A \succsim B.$$

First, we can show that only convex hulls should matter under Weak Order, Continuity and Mixture Independence.

Lemma 9.6 *When $\succsim$ satisfies Weak Order, Continuity and Mixture Independence, the following condition holds.*

Mixture Indifference*: For all $A \in \mathcal{Z}$,*

$$A \sim con A.$$

Proof First let us consider the case that A is a finite set, and let $A = \{x_1, \cdots, x_n\}$. We start with showing $\lambda A + (1-\lambda) con A = con A$ whenever $\lambda \leq 1/n$.

It is straightforward to see $\lambda A + (1-\lambda) con A \subset con A$. To show the converse, fix arbitrary $\lambda \leq 1/n$ and $x \in con A$. By the definition of a convex hull, there exist $\alpha_1, \cdots, \alpha_n \geq 0$ with $\sum_{k=1}^n \alpha_k = 1$ such that $x = \sum_{k=1}^n \alpha_k x_k$. Then for at least one j it must be that $\alpha_j \geq 1/n$, hence $\alpha_j \geq \lambda$. Now define $\widehat{\alpha}_1, \cdots, \widehat{\alpha}_n$ by

$$\widehat{\alpha}_j = \frac{\alpha_j - \lambda}{1-\lambda}$$

and

$$\widehat{\alpha}_k = \frac{\alpha_k}{1-\lambda}$$

for each $k \neq j$. Then

$$\sum_{k=1}^n \widehat{\alpha}_k = \frac{1}{1-\lambda}\left[\alpha_j - \lambda + \sum_{k\neq j} \alpha_j\right] = \frac{1}{1-\lambda}[1-\lambda] = 1.$$

Let $\widehat{x} = \sum_{k=1}^n \widehat{\alpha}_k x_k$, then $\widehat{x} \in con A$. Hence

$$\lambda x + (1-\lambda)\widehat{x} \in \lambda A + (1-\lambda) con A.$$

Notice that

$$\lambda x + (1-\lambda)\widehat{x} = \lambda x_j + (1-\lambda)\frac{\alpha_j - \lambda}{1-\lambda} x_j + (1-\lambda)\sum_{k\neq j} \frac{\alpha_k}{1-\lambda} x_k = \sum_{k=1}^n \alpha_k x_k = x.$$

Thus we obtain $x \in \lambda A + (1-\lambda) con A$. Hence $\lambda A + (1-\lambda) con A \supset con A$.

Now suppose $A \succ con A$. By Mixture Independence, for arbitrary $0 < \lambda \leq 1/n$,

$$\lambda A + (1-\lambda) con A \succ \lambda con A + (1-\lambda) con A = con A.$$

However, this contradicts $\lambda A + (1-\lambda) con A = con A$. Similarly when $A \prec con A$.

Let A be any compact set, then there is a sequence of finite sets $\{A^n\}$ that converges to it in the Hausdorff metric. Then, $\{con A^n\}$ converges to $con A$ in the Hausdorff metric as well.

Since $A^n \sim con A^n$ holds for all n, by Continuity we obtain $A \sim con A$. □

Theorem 9.3 *Preference* $\succsim$ *satisfies Weak Order, Continuity and Mixture Indifference if and only if it is represented by a continuous function* $U : \mathcal{Z} \to \mathbb{R}$ *and there exist a set* S *and a function* $u : \Delta(X) \times S \to \mathbb{R}$ *that is mixture-linear on* $\Delta(X)$, *and* *a function* $W : \prod_{s \in S} u(\Delta(X), s) \to \mathbb{R}$ *such that* U *has the form*

$$U(Z) = W\left(\left(\max_{a \in Z} u(a, s)\right)_{s \in S}\right).$$

Proof We show sufficiency of the axiom, as necessity is straightforward.

By Weak Order and Continuity, there is a continuous function $U : \mathcal{Z} \to \mathbb{R}$ which represents $\succsim$. By Mixture Indifference $U(Z) = U(con Z)$ holds for all $Z \in \mathcal{Z}$. Hence we consider U that is restricted to $\mathcal{K}$ hereafter.

Let S denote the unit sphere in $\mathbb{R}^X$. Let $C(S)$ denote the set of real-valued functions over S, which are bounded since S is compact.

Then, each $Z \in \mathcal{K}$ induces a support function $h_Z : S \to \mathbb{R}$ that is defined by

$$h_Z(s) = \max_{a \in Z} \langle a, s \rangle.$$

Let

$$C_{\mathcal{K}} = \{h_Z : Z \in \mathcal{K}\}$$

be the set of such support functions, then it is a subset of $C(S)$.

Also, when $g \in C_{\mathcal{K}}$ is given, we can define Z_g, a subset of $\Delta(X)$, by

$$Z_g = \bigcap_{s \in S} \{a \in \Delta(X) : \langle a, s \rangle \leq g(s)\},$$

which is an element of $\mathcal{K}$. Thus we have the following claim.

Claim For all $Y \in \mathcal{K}$ and $g \in C_{\mathcal{K}}$, $Z_{h_Y} = Y$ and $h_{Z_g} = g$. That is, h forms a one-to-one and onto mapping from $\mathcal{K}$ to $C_{\mathcal{K}}$.

Also, for all $Z, Z' \in \mathcal{K}$, if $Z \subset Z'$ then $h_Z \leq h_{Z'}$.

Now define $W : C_{\mathcal{K}} \to \mathbb{R}$ by

$$W(g) = U(Z_g).$$

By the above claim, W is well-defined.

Thus, for all $Z \in \mathcal{K}$ we have

$$U(Z) = W(h_Z) = W\left(\left(\max_{a \in Z} u(a, s)\right)_{s \in S}\right).$$

□

From the above proof strategy, a "subjective ex-post state" here is nothing but an expected utility preference, a (normalized) vector that evaluates lotteries.

When strengthening Mixture Indifference to Mixture Independence we can represent the preference by expected value of ex-post vNM utilities across "subjective states." Note, however, that the probability distribution showing up there has no meaning as subjective belief, since ex-post vNM utilities are dependent on the "subjective states."

Theorem 9.4 *$\succsim$ satisfies Weak Order, Continuity, Mixture Independence and Monotonicity if and only if there exist a mixture-linear and continuous function $U : \mathcal{Z} \to \mathbb{R}$ that represents $\succsim$ and a set S, a function $u : \Delta(X) \times S \to \mathbb{R}$ that is mixture-linear in $\Delta(X)$ for all $s \in S$, and a measure μ over S such that U has the form*

$$U(A) = \int_S \max_{a \in A} u(a, s) d\mu.$$

Proof Once we have the proof strategy, the rest is rather more about mathematical properties of $C_{\mathcal{K}}$ in the space $C(S)$. Thus we give a sketch of the sufficiency part, as the necessity of them is transparent.

Since $\succsim$ satisfies Weak Order, Continuity and Mixture Independence, by following the same argument for the proof of Theorem 4.4, we can establish existence of a mixture-linear and continuous representation $U : \mathcal{Z} \to \mathbb{R}$. Since Mixture Indifference is met, we can limit attention to the restriction of U to $\mathcal{K}$.

Then, by the same argument as in the proof of Theorem 9.3, we can define a function $I : C_{\mathcal{K}} \to \mathbb{R}$ by

$$I(g) = U(Z_g)$$

and we have

$$U(Z) = I(h_Z).$$

Note that the support-function description is preserved under mixtures. That is, for all $g, g' \in C_{\mathcal{K}}$ and $\lambda \in [0, 1]$,

$$Z_{\lambda g + (1-\lambda) g'} = \lambda Z_g + (1 - \lambda) Z_{g'}.$$

Then, by mixture-linearity of U,

$$\begin{aligned} I(\lambda g + (1 - \lambda) g') &= U(Z_{\lambda g + (1-\lambda) g'}) \\ &= U(\lambda Z_g + (1 - \lambda) Z_{g'}) \\ &= \lambda U(Z_g) + (1 - \lambda) U(Z_{g'}) \\ &= \lambda I(g) + (1 - \lambda) I(g') \end{aligned}$$

Thus I is a mixture-linear function.

We extend I from $C_{\mathcal{K}}$ to the entire $C(S)$. On $C_{\mathcal{K}} - C_{\mathcal{K}}$, for each $g \in C_{\mathcal{K}} - C_{\mathcal{K}}$, where $g', g'' \in C_{\mathcal{K}}$ give $g = g' - g''$, we define

$$I(g) = I(g') - I(g'').$$

Then I is a linear and continuous real-valued function over $C_{\mathcal{K}} - C_{\mathcal{K}}$.

One can show that $C_{\mathcal{K}} - C_{\mathcal{K}}$ is topologically dense in $C(S)$. Hence we can extend I to the entire $C(S)$ continuously, and it is linear.

Because $f \leq \|f\|\mathbf{1}$ holds for all $f \in C(S)$, where $\|\cdot\|$ denotes the uniform norm on $C(S)$ and $\mathbf{1}$ denotes the constant function giving 1, by Monotonicity we have $I(f) \leq I(\|f\|\mathbf{1}) = \|f\|I(\mathbf{1})$. Thus the function I is Lipschitz continuous with respect to constant $I(\mathbf{1})$, hence it is bounded. See the supplement paper by Dekel et al. [25] about the issue of why Monotonicity is needed.

Now by the Riesz representation theorem (Theorem 14.12 in Aliprantis and Border [3]), there is a measure over S, denoted by μ, so that

$$I(g) = \int_S g(v)\mu(dv).$$

Now restrict this to the subspace $C_{\mathcal{K}}$, then for all $Z \in \mathcal{K}$ we have

$$U(Z) = I(h_Z) = \int_S \max_{a \in Z} \langle a, v \rangle \mu(dv).$$

□

9.3.1 *Dynamic Extension?*

One might want to think of a dynamic extension of the model of subjective states above, in which the decision maker learns realization of subjective states. But it is not easy, because we cannot simply take a preference ranking conditioned by a subjective state as a part of the primitive, because it is subjective and unobservable—if we could observe it we would not need the model of subjective states!

Still we can find a role of subjective states in dynamic environments, by trying to identify how subjective states *supplement* the belief process over objective ones. See for example Takeoka [96], Dillenberger et al. [27] for this type of approach.

Chapter 10
Information and Knowledge

Since this book is not about game theory or information economics, logical reasoning about uncertainty (especially the strategic one) is not in the main scope. Information and knowledge, however, are surely indispensable factors in decision making, and let me cover these as far as they are in the scope of the book. For game theoretic treatment of knowledge, the readers should consult notable textbooks such as Osborne and Rubinstein [85], Fudenberg and Tirole [37].

Here we proceed based on the assumption that we can directly observe the decision maker's cognitive process/status about "knowing" something, instead of adopting the revealed preference approach. Let me call it the *cognitive approach*.

The cognitive approach does not exclude the revealed preference approach. For example, the fact that the decision maker "knows that some event has happened" may be revealed from betting choices, when it is interpreted as assigning probability 1 to such an event. But we illustrate the cognitive approach as it is, since it will be intuitive by itself as well.

10.1 Possibility Correspondence

Denote the set of states by Ω. Consider a possibility correspondence $P : \Omega \to 2^{\Omega} \setminus \{\emptyset\}$, which is interpreted as saying

> if the true state is ω, the decision maker knows that the true state is in $P(\omega)$ but does not know which one in it is true.

In this sense a possibility correspondence describes the decision maker's information/knowledge structure.

"Rational" possibility correspondence P is supposed to satisfy the following axioms:

P1 (Knowledge): For all $\omega \in \Omega$, $\omega \in P(\omega)$.

T. Hayashi, *Decision Theory*, Monographs in Mathematical Economics 9,
https://doi.org/10.1007/978-981-95-2200-2_10

P2 (Know that you know): For all $\omega, \omega' \in \Omega$, if $\omega' \in P(\omega)$ then $P(\omega') \subset P(\omega)$.
P3 (Know that you don't know): For all $\omega, \omega' \in \Omega$, if $\omega' \in P(\omega)$ then $P(\omega') \supset P(\omega)$.

Why are P2 and P3 named as above? It will be better to explain this after introducing the knowledge operator model, another model of knowledge.

Remark 10.1 Note that P3 implies P2 under P1 (Geanakoplos [40]). To see this, for any $\omega \in \Omega$ and $\omega' \in P(\omega)$, pick any $\omega'' \in P(\omega')$. Then from P3 we have $P(\omega') \supset P(\omega)$ and $P(\omega'') \supset P(\omega')$, which imply $P(\omega'') \supset P(\omega)$.

On the other hand, from P1 we have $\omega \in P(\omega)$, the above-obtained $P(\omega'') \supset P(\omega)$ yields $\omega \in P(\omega'')$.

Then from P3 we have $P(\omega) \supset P(\omega'')$. Since $\omega'' \in P(\omega'')$ because of P1, $P(\omega) \supset P(\omega'')$ delivers $\omega'' \in P(\omega)$.

To see the points of the axioms, consider the following examples.

Example 10.1 Let $\Omega = \{a, b, c\}$, and consider a possibility correspondence given by $P(a) = \{b\}$, $P(b) = P(c) = \{b, c\}$. Then, despite $b \in P(a)$ we have $P(b) = \{b, c\} \supsetneq \{b\} = P(a)$, which violates P1 and P2 while it satisfies P3.

Note that we cannot given an example that violates P2 alone, since P3 implies P2 under P1.

Example 10.2 Denote "a dog barks" by b and "a dog does not bark" by n, and let $\Omega = \{b, n\}$. Then let $P(b) = \{b\}$, meaning that once a dog barks we surely hear it.

Now consider that we have $P(n) = \{b, n\}$, because of having the possibility in mind that a dog actually barks and we fail to hear it.

Then, despite $b \in P(n)$ we have $P(b) \not\supset P(n)$, and we have a violation of P3. For, from $P(b) = \{b\}$ we see that once a dog barks we surely hear it and should be able to exclude the possibility that a dog actually barks and we fail to hear it.

In order that P3 is met, we have to modify either as $P(n) = \{n\}$ or $P(b) = \{b, n\}$, where the latter requires us to refute the assumption that once a dog barks we surely hear it.

Definition 10.1 Say that a possibility correspondence P is partitional if there is a partition of Ω such that for all $\omega \in \Omega$ the set $P(\omega)$ is the element of the partition that contains ω.

The above two examples are not partitional. Also,

$$\Omega = \{a, b, c\}, \quad P(a) = \{a, b\}, \quad P(b) = P(c) = \{b, c\}$$

is not partitional but

$$\Omega = \{a, b, c\}, \quad P(a) = \{a\}, \quad P(b) = P(c) = \{b, c\}$$

is partitional.

Proposition 10.1 *A possibility correspondence satisfies P1-P3 if and only if it is partitional.*

Proof Necessity of P1-P3 is straightforward. We show sufficiency.

From P2 and P3 we obtain

For all $\omega, \omega' \in \Omega$, if $\omega' \in P(\omega)$ then $P(\omega') = P(\omega)$.

Now, for any $\omega, \omega' \in \Omega$, suppose $P(\omega)$ and $P(\omega')$ intersect and pick any $\omega'' \in P(\omega) \cap P(\omega')$. Then from above we obtain $P(\omega) = P(\omega') = P(\omega'')$.

Also, from P1, $\bigcup_{\omega \in \Omega} P(\omega) = \Omega$. Thus P is partitional. □

10.2 Knowledge Operator

Given a possibility correspondence P, one can formalize the statement "the decision maker knows event E at state ω" by

$$P(\omega) \subset E,$$

that is, by saying that he recognizes E to be true at ω.

Thus, given P, we can define a operator $K : 2^{\Omega} \to 2^{\Omega}$ by

$$K(E) = \{\omega \in \Omega : P(\omega) \subset E\}. \tag{10.1}$$

That is, $K(E)$ refers to the set of states at which event E is recognized as true. This allows us to equivalently formalize the statement "the decision maker knows event E at state ω" by

$$\omega \in K(E).$$

Whether P is partitional or not, the following three properties follow from the definition:

K1: $K(\Omega) = \Omega$.
K2: If $E \subset F$ then $K(E) \subset K(F)$.
K3: $K(E) \cap K(F) = K(E \cap F)$.

Suppose P satisfies P1. Then, if $\omega \in K(E)$, from (10.1) we have $P(\omega) \subset E$. Since P1 says $\omega \in P(\omega)$, we obtain $\omega \in E$. Since $\omega \in K(E)$ was arbitrary, we obtain $K(E) \subset E$.

Hence P1 for P implies the following property for K.

K4 (Knowledge): $K(E) \subset E$.

Suppose P satisfies P2. Then, if $\omega \in K(E)$, from (10.1) we have $P(\omega) \subset E$. Now pick any $\omega' \in P(\omega)$, then from P2 it follows that $P(\omega') \subset P(\omega)$, which implies $P(\omega') \subset E$. Hence $\omega' \in K(E)$. Since $\omega' \in P(\omega)$ was arbitrary, we obtain $P(\omega) \subset K(E)$, that is, $\omega \in K(K(E))$. Since $\omega \in K(E)$ was arbitrary, we obtain $K(E) \subset K(K(E))$.

Hence P2 for P implies the following property for K:

K5 (Know that you know): $K(E) \subset K(K(E))$.

Now it is clear why the property is named in this way: if the decision maker knows E at ω, he knows $K(E)$ at ω, that is, at ω he knows that he knows E.

Suppose P satisfies P3. Then, if $\omega \in \Omega \setminus K(E)$, from (10.1) we have $P(\omega) \not\subset E$. Now pick any $\omega' \in P(\omega)$, then from P3 it follows that $P(\omega') \supset P(\omega)$, which implies $P(\omega') \not\subset E$. Hence $\omega' \in \Omega \setminus K(E)$. Since $\omega' \in P(\omega)$ was arbitrary, we obtain $P(\omega) \subset \Omega \setminus K(E)$, that is, $\omega \in K(\Omega \setminus K(E))$. Since $\omega \in \Omega \setminus K(E)$was arbitrary, we obtain $\Omega \setminus K(E) \subset K(\Omega \setminus K(E))$.

Hence P3 for P implies the following property for K:

K6 (Know that you don't know): $\Omega \setminus K(E) \subset K(\Omega \setminus K(E))$.

Now it is clear why the property is named in this way: if the decision maker does not know E at ω, he knows $K(E)(\Omega \setminus K(E))$ at ω, that is, at ω he knows that he does not know E.

Here we started from a possibility correspondence P and define the corresponding knowledge operator K. We can do the reverse argument as well. Given a knowledge operator K, we can define the correspondence P by

$$P(\omega) = \bigcap_{E:K(E)\ni\omega} E. \tag{10.2}$$

We show K4 for K implies P1 for P. Given any $\omega \in \Omega$, from K1, any E satisfying $\omega \in K(E)$ must satisfy $\omega \in E$. Hence $\omega \in \bigcap_{E:K(E)\ni\omega} E$, that is, $\omega \in P(\omega)$.

We show K5 for K implies P2 for P. Let P be defined from K by (10.2). Pick any $\omega \in \Omega$ and any $\omega' \in P(\omega)$. Pick $\omega'' \in P(\omega')$.

Thus, whenever $\omega \in K(E')$, from K5, $\omega \in KK(E')$. From (10.2) and $\omega' \in P(\omega)$ it follows that $\omega' \in K(E')$ for any such case. From (10.2) and $\omega'' \in P(\omega')$ it follows that $\omega'' \in E'$ for any such case.

Thus we obtain that whenever $\omega \in K(E')$ holds it follows $\omega'' \in E'$. From (10.2) it follows that $\omega'' \in P(\omega)$. Thus we have shown $P(\omega') \subset P(\omega)$.

Finally, we show K6 for K implies P3 for P. Pick any $\omega \in \Omega$ and pick any $\omega' \in P(\omega)$.

(a) Then from (10.2), $\omega' \in E'$ whenever $\omega \in K(E')$.
Suppose $P(\omega') \not\supset P(\omega)$. Then there exists $\omega'' \in P(\omega)$ such that $\omega'' \notin P(\omega')$.
(b) Since $\omega'' \in P(\omega)$, from (10.2), $\omega'' \in E'$ whenever $\omega \in K(E')$.
(c) Since $\omega'' \notin P(\omega')$, from (10.2) there exists E such that $\omega'' \notin E$ while $\omega' \in K(E)$.

Case 1 Suppose $\omega \in K(E)$. Then from (b) we obtain $\omega'' \in E$, which is a contradiction to (c).

Case 2 Suppose $\omega \notin K(E)$. Then by K6 $\omega \in K(\Omega \setminus K(E))$. By (a) we obtain $\omega' \in \Omega \setminus K(E)$, which is a contradiction to (c).

10.3 Common Knowledge

Let us consider a two-person case, in which each agent is denoted by $i = 1, 2$, which can be extended to n persons. For each $i = 1, 2$, denote his knowledge operator by K_i. Then, consider a chain of arguments given by:

- if $\omega \in K_1(E)$, then at state ω agent 1 knows event E,
- if $\omega \in K_2(K_1(E))$, then at state ω agent 2 knows that agent 1 knows event E,
- if $\omega \in K_1(K_2(K_1(E)))$, then at state ω agent 1 knows that agent 2 knows that 1 knows event E,
- and so on.

Now, say that event E is common knowledge at state ω if

$$\omega \in \bigcap_{n=1}^{\infty} \bigcap_{(i_1,\cdots,i_n)\in\{1,2\}^n} K_{i_1} \cdots K_{i_n}(E) \equiv C(E).$$

For example, let $\Omega = \{\omega_1, \omega_2, \omega_3, \omega_4, \omega_5, \omega_6, \omega_7, \omega_8\}$ and consider that each individual's information structure is given by

$$P_1 = \{\{\omega_1, \omega_2\}, \{\omega_3, \omega_4, \omega_5\}, \{\omega_6\}, \{\omega_7, \omega_8\}\},$$

$$P_2 = \{\{\omega_1\}, \{\omega_2, \omega_3, \omega_4\}, \{\omega_5\}, \{\omega_6, \omega_7\}, \{\omega_8\}\}.$$

Note that P_1 and P_2 are partitional.

Consider event $E = \{\omega_1, \omega_2, \omega_3, \omega_4\}$, then it cannot be common knowledge, since

$$K_1(E) = \{\omega_1, \omega_2\}, \quad K_2(E) = \{\omega_1, \omega_2, \omega_3, \omega_4\},$$

$$K_2K_1(E) = \{\omega_1\}, \quad K_1K_2(E) = \{\omega_1, \omega_2\},$$

$$K_1K_2K_1(E) = \emptyset, \quad K_2K_1K_2(E) = \{\omega_1\}.$$

On the other hand, event $F = \{\omega_1, \omega_2, \omega_3, \omega_4, \omega_5\}$ is common knowledge at any state in it. Since

$$K_1(F) = F, \ K_2(F) = F$$

it follows that

$$K_2K_1(F) = F, \quad K_1K_2(F) = F,$$

$$K_1K_2K_1(F) = F, \quad K_2K_1K_2(F) = F,$$

and we see

$$C(F) = F.$$

An event F is said to be **self-evident between 1 and 2** if $P_i(\omega) \subset F$ holds for every $i = 1, 2$ for all $\omega \in F$.

The lemma and proposition below follow from the definition.

Lemma 10.1 *The following three conditions are equivalent:*

(a): $K_i(F) = F, i = 1, 2.$
(b): *F is self-evident between 1 and 2.*
(c): *For each i* $= 1, 2$, *the set of obtained as a union of elements of* P_i.

Proof Assume (a). Then for any $\omega \in F$, $\omega \in K_i(F)$, that is, $P_i(\omega) \subset F$, for each i. Hence (b) is met.

Assume (b). Then $F = \bigcup_{\omega \in F} P_i(\omega)$ for each i, hence (c) is met.

Assume (c), then $F = \bigcup_{\omega \in F} P_i(\omega)$ for each i. Then $P_i(\omega) \subset F$ for all $\omega \in F$, that is, $F \subset K_i(F)$.

$K_i(F) \subset F$ follows from K4. This is true for each i. □

Proposition 10.2 *Assume that* P_1, P_2 *are partitional. Then, event* E *is common knowledge at state* ω *if and only if there is a self-evident set* F *such that* $\omega \in F \subset E$.

Proof **"Only If" Part** Since

$$E \supset K_iE \supset K_jK_iE \supset K_iK_jK_iE \supset \cdots \ni \omega$$

for either $i \in \{1, 2\}$ and $j \neq i$, there is

$$F_i = K_iK_jK_i \cdots K_iE$$

with

$$K_jF_i = F_i.$$

By K5,

$$F_i = K_i K_j K_i \cdots K_i E \subset K_i K_i K_j K_i \cdots K_i E = K_i F_i.$$

By K4,

$$K_i F_i \subset F_i.$$

Hence

$$K_i F_i = F_i.$$

"If" Part Suppose there is F with $\omega \in F \subset E$ such that F is self-evident.

Then by the previous lemma, every set with the form $K_i K_j K_i \cdots K_i F$ coincides with F.

From K2, ω belongs to every set with the form $K_i K_j K_i \cdots K_i E$. □

10.4 Can People Agree to Disagree?

When you receive some information saying that the price of a stock is going up. You go to the market to buy the stock, and you meet a trader who is going to sell it. If you are "rational" in the sense that you are not stubborn, you should be surprised and wonder, "why is this guy willing to sell?" and you revise your belief. Similarly for the seller. If he is "rational" in the sense that he is not stubborn, he should be surprised and wonder, "why is this guy willing to buy?" and he revise his belief.

Eventually, after such contemplation and speculation, you will say "OK, I won't buy," and he will say "OK, I won't sell." Let me make it clear that the "rationality" assumption is that both have the same prior belief at the fundamental level and any difference in belief comes only from difference in information they receive.

Theorem 10.1 (Aumann [9]) *Let Ω be a finite set, and assume that P_1, P_2 are partitional, and that they have a common prior μ. Suppose it is common knowledge at state ω that 1's posterior probability of event E is η_1 and 2's posterior probability of E is η_2. Then $\eta_1 = \eta_2$.*

Proof From Proposition 10.2, there is a self-evident set F such that

$$\omega \in F \subset \{\omega \in \Omega : \mu(E|P_1(\omega)) = \eta_1,\ \mu(E|P_2(\omega)) = \eta_2\}.$$

From Lemma 10.1, we have mutually disjoint elements of P_1, say denoted by $P_{11}, \cdots, P_{1m}$, and mutually disjoint elements of P_2, say denoted by $P_{21}, \cdots, P_{2n}$, such that

$$F = \bigcup_{k=1}^{m} P_{1k} = \bigcup_{j=1}^{n} P_{2j}$$

and

$$\mu(E|P_{1k}) = \eta_1, \ \mu(E|P_{2j}) = \eta_2$$

for each $k = 1, \cdots, m$ and each $j = 1, \cdots, n$.

Since

$$\mu(E|F)\mu(F) = \mu(E \cap F),$$

it follows that

$$\begin{aligned} \mu(E|F)\sum_{k=1}^{m} \mu(P_{1k}) &= \sum_{k=1}^{m} \mu(E \cap P_{1k}) \\ &= \sum_{k=1}^{m} \mu(E|P_{1k})\mu(P_{1k}) \\ &= \eta_1 \sum_{k=1}^{m} \mu(P_{1k}), \end{aligned}$$

which yields $\mu(E|F) = \eta_1$. Likewise, we obtain $\mu(E|F) = \eta_2$, which delivers $\eta_1 = \eta_2$. □

The statement below follows immediately.

Corollary 10.1 *Suppose Ω is finite and P_1, P_2 are partitional, and the agents have a common prior. Let L be a bet. Then it cannot be common knowledge that 1 believes the conditional expected value of bet L is above α and 2 believes it is below α.*

10.5 Can We Model Unawareness?

In our natural language, a statement like

> I am unaware of something, and I do not know what I am unaware of, but I am unaware that I am unaware of something

is quite intuitive. It is difficult to model this, however.

Dekel et al. [23] showed that it is impossible to model unawareness in a meaningful way, as far as we adopt the standard state space model.

Let Ω be the set of states, let $K : 2^{\Omega} \to 2^{\Omega}$ denote the knowledge operator, and let $U : 2^{\Omega} \to 2^{\Omega}$ denote the unawareness operator which we want to think of, and now consider the triple (Ω, K, U).

The knowledge operator is supposed to meet the two natural requirements,

K1 (Necessitation): $K(\Omega) = \Omega$.
K2 (Monotonicity): If $E \subset F$ then $K(E) \subset K(F)$.

Given an unawareness operator U, the statement

$$\omega \in U(E)$$

is interpreted as saying that at ω the decision maker is unaware of event E.

According to what usually mean by being "unaware," it will be natural to assume the following three conditions for U:

Plausibility: For all E, $U(E) \subset \neg K(E) \cap \neg K \neg K(E)$.
KU Introspection: For all E, $KU(E) = \emptyset$.
AU Introspection: For all E, $U(E) \subset U(U(E))$.

Plausibility says that if you are unaware of an event E then you don't know E and you don't know that you don't know E, as either "knowing E" or "knowing that you don't know E" sounds possible only if you are already aware of such an E. KU Introspection says it is impossible for you to know that you are unaware of an event E. It is funny if we say you know that you are unaware of an event E, since it seems possible only if you are already aware of such an E. AU Introspection says that if you are unaware of an event E you must be unaware that you are unaware of E. It is funny again if we say you are aware that you are unaware of an event E when you are unaware of E in fact, since it seems possible only if you are already aware of such an E.

Theorem 10.2 *Suppose that (Ω, K, U) satisfies Plausibility, KU Introspection and AU introspection. Then:*

(i) If K satisfies Necessitation, then $U(E) = \emptyset$ holds for all E.
(ii) If K satisfies Monotonicity, then $U(E) \subset \neg K(F)$ holds for all E and F.

Proof From AU Introspection and Plausibility, we have

$$U(E) \subset U(U(E)) \subset \neg K \neg K(U(E)).$$

Since KU Introspection means $\neg KU(E) = \Omega$, we obtain $U(E) \subset \neg K(\Omega)$.

Under Necessitation, from $K(\Omega) = \Omega$ it follows that $U(E) = \emptyset$.

Under Monotonicity, for any event F, $K(F) \subset K(\Omega)$ holds, implying $\neg K(F) \supset \neg K(\Omega)$. Thus we obtain $U(E) \subset \neg K(F)$. □

The first conclusion is that you cannot be unaware of anything. The second one says that if you are unaware of something you cannot know anything.

The above result shows that it is impossible to model unawareness in a meaningful way, as far as we consider state space in the abstract way. Then, how can we model the concept of unawareness which we have in mind as intuition? I cannot cover all the related developments, but let me just mention the basic direction.

The direction is to think of a certain structure (especially a product structure) of a state space, instead of abstract state space without any structure. For example, let us denote the set of objects by Z and let S_z denote the set of states regarding $z \in Z$. Then, if the decision maker is not fully aware of entire Z but aware only of its subset $Y \subset Z$, the set $S_Y = \prod_{z \in Y} S_z$ is seen as the set of states which he is aware of. See Heifetz et al. [61] and Li [75].

Such development motivates a question about how a decision maker does or should extend her belief when her awareness grows and faces an "expanded" state space. Karni and Viero [66] propose so-called "reverse Bayesianism," which is mathematically seen as a version of Kolmogorov extension to a larger state space such that the marginal of the extended belief onto the original state space coincides with the original one.

Chapter 11
Social Decision Making Under Uncertainty

The Pareto principle saying "if everybody says X is better than Y so should the society" sounds universal when it is told in an abstract setting. Opposing it will even sound "undemocratic." It is not obvious, however, under uncertainty. Let us see this below.

From now on, denote the set of individuals by I, which is finite, and denote the society by 0. Let $\mathbb{R}^I$ denote the vector space as the set of mappings from I to $\mathbb{R}$. Let $\mathbf{0}_I$ denote the zero vector that assigns 0 to every $i \in I$, and let $\mathbf{1}_I$ denote the vector that assigns 1 to every $i \in I$. Also, let $\mathbb{R}^{\{0\} \cup I}$ denote the vector space as the set of mappings from $\{0\} \cup I$ to $\mathbb{R}$. Let $\mathbf{0}_{\{0\} \cup I}$ denote the zero vector that assigns 0 to every $i \in \{0\} \cup I$, and let $\mathbf{1}_{\{0\} \cup I}$ denote the vector that assigns 1 to every $i \in \{0\} \cup I$.

11.1 Social Decision Making Under Risk

First, consider the setting that probability distributions over social outcomes are given as *objects*, and how the society should rank between them.

11.1.1 The Harsanyi Aggregation Theorem

Let X denote the set of social outcomes, which is assumed to be finite for simplicity. Then, denote the set of probability distributions over X by $\Delta(X)$.

For each $i \in I$, denote his preference held over $\Delta(X)$ by $\succsim_i$. Also, denote the society's normative ranking over $\Delta(X)$ by $\succsim_0$. Note that they not only express which outcomes are better or worse, but also which risky prospects are better or worse, hence they express individuals' actual risk attitudes and the society's normative risk attitudes.

T. Hayashi, *Decision Theory*, Monographs in Mathematical Economics 9,
https://doi.org/10.1007/978-981-95-2200-2_11

Below, we assume that each individual's risk preference $\succsim_i$ satisfies the expected utility theory. This is a *descriptive assumption*. We also assume that the society's ranking $\succsim_0$ satisfies the expected utility theory. This is a *normative requirement*.

Why should the social ranking satisfy the expected utility theory? It will be obvious for completeness and transitivity, while there is a view that we should dare to allow incompleteness. Continuity is mostly a technical requirement, but it has a normative implication. It says that any kind of risk, if it is sufficiently small, should be acceptable under sufficient compensation. This could be a significant requirement. See Fishburn [36] for a study on expected utility theory without continuity. Let us assume continuity, however.

The independence axiom is most problematic in descriptive argument, but as a normative requirement it is understandable because it is equivalent to dynamic consistency. Note of course that such equivalence is under consequentialism and timing indifference, as we saw in previous chapters. We come back to this point later.

Next, we consider a Pareto principle regarding choice over lotteries. In an abstract setting, the statement "if everybody says X is better than Y so should the society" sounds innocuous. It is not any longer on the domain with a structure, which is risk/uncertainty here. Applying the Pareto principle to ranking over lotteries means that every individual is taken to be responsible for his risk attitude as a "matter of taste," and it has to be respected in determining the social ranking. To emphasize this point, the Pareto principle being applied under risk/uncertainty in such a way is called the **ex-ante** Pareto principle.

For example, consider two agents, A and B, who are extremely risk-loving. Consider a bet in which A pays 1000 dollars to B with probability half and B pays 1000 dollars to A with probability half. Compare the bet with an option that they do nothing, then both A and B will prefer to bet. Applying the ex-ante Pareto principle here means that there is no reason to object to it when both A and B prefer to bet. Note again that such application requires the presumption that risk attitudes are a "matter of taste" and each individual is perfectly responsible for his own.

Axiom 11.1 (Ex-Ante Weak Pareto) if $l \succ_i l'$ for all $i \in I$, then $l \succ_0 l'$.

Axiom 11.2 (Ex-Ante Strong Pareto) if $l \succsim_i l'$ for all $i \in I$ and $l \succ_i l'$ for at least one $i \in I$, then $l \succ_0 l'$.

We assume the minimal agreement condition, which is natural as everybody will dislike the outcome like "the world becomes extinct."

Minimal Agreement Condition: there exist c, c' such that $c \succ_i c'$ holds for every $i \in I$.

The statement below was first shown by Harsanyi [55], while we follow the proof by De Meyer and Mongin [20].

Theorem 11.1 *Suppose* $\succsim_i$ *satisfies the expected utility theory for every* $i \in I$, *and that* $\succsim_0$ *satisfies the theory as well. For each* $i \in I$, *fix a function* $u_i : X \to \mathbb{R}$ *that forms an expected utility representation of* $\succsim_i$, *and fix a function* $u_0 : X \to$

$\mathbb{R}$ that forms an expected utility representation of $\succsim_0$. Also, assume the Minimal Agreement Condition. Then:

(i) $(\succsim_i)_{i\in I\cup\{0\}}$ *satisfy Ex-ante Weak Pareto if and only if there exist a vector* $\lambda \in \mathbb{R}^I_+ \setminus \{\mathbf{0}\}$ *and a constant* μ_0 *such that*

$$u_0 = \sum_{i\in I} \lambda_i u_i + \mu_0$$

holds

(ii) $(\succsim_i)_{i\in I\cup\{0\}}$ *satisfy Ex-ante Strong Pareto if and only if there exist a vector* $\lambda \in \mathbb{R}^I_{++}$ *and a constant* μ_0 *such that*

$$u_0 = \sum_{i\in I} \lambda_i u_i + \mu_0$$

holds.

Proof (i) Let

$$C = \{(u_0 l - u_0 l', (u_i l - u_i l')_{i\in I}) \in \mathbb{R}^{\{0\}\cup I} : l, l' \in \Delta(X)\},$$

then it is a convex set. From Ex-ante Weak Pareto,

$$C \cap \left((-\mathbb{R}_+) \times \mathbb{R}^I_{++}\right) = \emptyset.$$

Hence by the separating hyperplane theorem there exist a vector $q \in \mathbb{R}^{\{0\}\cup I} \setminus \{\mathbf{0}_{\{0\}\cup I}\}$ and a number α such that

$$qv \geq \alpha \geq qw, \quad \forall v \in C, \ \forall w \in (-\mathbb{R}_+) \times \mathbb{R}^I_{++}$$

holds.

Since every $v \in C$ satisfies $-v \in C$, $qv \geq \alpha$ and $q(-v) \geq \alpha$, which implies $\alpha = 0$ and

$$qv = 0, \quad \forall v \in C.$$

Thus, for all $w \in (-\mathbb{R}_+) \times \mathbb{R}^I_{++}$, $0 \geq qw$. Now suppose $q_i > 0$ for some $i \in I$. Then, by taking $w_i = 1$, $w_0 = -\varepsilon$ and $w_j = \varepsilon$ for all $j \in I \setminus \{i\}$ with sufficiently small $\varepsilon > 0$, we obtain

$$qw = q_i - \varepsilon\left(q_0 + \sum_{j\in I\setminus\{i\}} q_j\right) > 0,$$

which is a contradiction to the separation condition. Hence $q_i \leq 0$ for all $i \in I$.

Suppose $q_0 < 0$. Then, by taking $w_0 = -1$ and taking $w_i = \varepsilon$ for all $i \in I$ with sufficiently small $\varepsilon > 0$, we have

$$qw = -q_0 + \varepsilon \sum_{i \in I} q_i > 0,$$

which contradicts the separation condition. Hence $q_0 \geq 0$.

Also, suppose $q_0 = 0$. Since $q \in \mathbb{R}^{\{0\} \cup I} \setminus \{\mathbf{0}_{\{0\} \cup I}\}$ it must be that $q_I \in -\mathbb{R}^I_+ \setminus \{\mathbf{0}_I\}$. Since $qv = 0$ holds for all $v \in C$, we have

$$0 = \sum_{i \in I} q_i(u_i l - u_i l')$$

for all $l, l' \in \Delta(X)$. However, from the Minimal Agreement Condition there exist $l, l' \in \Delta(X)$ such that $u_i l - u_i l'$ for all $i \in I$, which results in a contradiction. Hence $q_0 > 0$.

Now suppose $q_i = 0$ for all $i \in I$. Then we have

$$0 = q_0(u_0 l - u_0 l')$$

for all $l, l' \in \Delta(X)$, but this again cannot be compatible with the Minimal Agreement Condition and Ex-ante Weak Pareto.

Summing up, there exist $q_0 > 0$ and $q_I \in -\mathbb{R}^I_+ \setminus \{\mathbf{0}_I\}$ such that

$$q_0(u_0 l - u_0 l') = \sum_{i \in I} -q_i(u_i l - u_i l')$$

for all $l, l' \in \Delta(X)$. Finally, for each $i \in I$, let $\lambda_i = -q_i/q_0$. Fix some $l' \in \Delta(X)$ and let $\mu_0 = u_0 l' + \sum_{i \in I} -(q_i/q_0) u_i l'$, then we obtain

$$u_0 = \sum_{i \in I} \lambda_i u_i + \mu_0.$$

(ii) Let

$$C = \{(u_0 l - u_0 l', (u_i l - u_i l')_{i \in I}) \in \mathbb{R}^{\{0\} \cup I} : l, l' \in \Delta(X)\},$$

then it is a convex set. From Ex-ante Strong Pareto,

$$C \cap \Big((-\mathbb{R}_+) \times (\mathbb{R}^I_+ \setminus \{\mathbf{0}_I\})\Big) = \emptyset.$$

Hence by the separating hyperplane theorem (stronger version) there exist a vector $q \in \mathbb{R}^{\{0\}\cup I} \setminus \{\mathbf{0}_{\{0\}\cup I}\}$ and a number α such that

$$qv \geq \alpha > qw, \quad \forall v \in C, \ \forall w \in (-\mathbb{R}_+) \times (\mathbb{R}^I_+ \setminus \{\mathbf{0}_I\})$$

holds.

Since every $v \in C$ satisfies $-v \in C$, $qv \geq \alpha$ and $q(-v) \geq \alpha$, which implies $\alpha = 0$ and

$$qv = 0, \quad \forall v \in C.$$

Thus, for all $w \in (-\mathbb{R}_+) \times (\mathbb{R}^I_+ \setminus \{\mathbf{0}_I\})$ it holds $0 > qw$. Now suppose $q_i \geq 0$ for some $i \in I$. Then, by taking $w_i = 1$, $w_0 = 0$ and $w_j = 0$ for all $j \in I \setminus \{i\}$ we have

$$qw = q_i \geq 0,$$

which is a contradiction to the separating condition. Hence we have $q_i < 0$ for all $i \in I$.

The rest is the same as in (i). □

Remark 11.1 The welfare weights obtained in the above theorem are dependent on **how we fix each individual's expected utility representation**, and has no meaning beyond that. For example, suppose we fix u_i for $\succsim_i$ and u_j for $\succsim_j$, and fix u_0 for $\succsim_0$, and that they are related in the form

$$u_0 = u_i + u_j,$$

that is, the welfare weight vector is given by $(1, 1)$.

However, if we consider a function that gives a different expected utility representation of $\succsim_i$, let's say $\widetilde{u}_i = 2u_i$, then we have

$$u_0 = u_i + u_j = \frac{1}{2} \cdot 2u_i + u_j = \frac{1}{2} \cdot \widetilde{u}_i + u_j$$

and the corresponding welfare vector is $\left(\frac{1}{2}, 1\right)$.

Once you change the scale of an individual's expected utility representation of his risk preference you obtain a different weight vector. Thus, unless you have a determinate reason about which is the "true" one among arbitrarily many expected utility representations of the same risk preference, the welfare weight vector obtained above has no ethical meaning.

Remark 11.2 Also note that even when the social ranking follows the expected utility theory the aggregation method does not have to be additive when you consider its representations *other than the expected utility ones*. The expected utility theory says only that a preference satisfying its set of axioms allows an expected utility representation, and not that its representation has to be of the expected utility form. What do I mean?

The above theorem implies that for a list of *whole* representations of preferences over lotteries

$$U_i(p) = \sum_{x \in S(p)} u_i(x)p(x), \quad i \in \{0\} \cup I,$$

we have

$$U_0(p) = \sum_{i \in I} \lambda_i U_i(p).$$

However, by taking arbitrary monotone transformation f_i, the function defined by

$$f_i(U_i(p))$$

represents the same risk preference $\succsim_i$, while not necessarily in an expected utility form.

For example, taking exponential transformation and having

$$e^{U_0(p)} = e^{\sum_{i \in I} \lambda_i U_i(p)} = \prod_{i \in I} \left(e^{U_i(p)}\right)^{\lambda_i}$$

does not change either of each individual's risk preference or the social ranking. Now by rewriting

$$V_i(p) = e^{U_i(p)}, \quad i \in \{0\} \cup I$$

we obtain

$$V_0 = \prod_{i \in I} (V_i(p))^{\lambda_i},$$

which is a multiplicative aggregation. This is the key to the so-called Harsanyi-Sen debate about utilitarianism, which is nicely illustrated in Weymark [100].

11.1.2 Are Risk Attitudes a "Matter of Taste?"

The ex-ante Pareto condition requires that each individual is perfectly responsible for his own risk attitude, not just his preference over outcomes, as a "matter of taste."

This is not an obvious requirement. Thus let us think of a weaker Pareto condition, in which each individual is responsible only for his preference over outcomes (see Hammond [54] for further discussions).

Axiom 11.3 (Ex-Post Weak Pareto) if $\delta_x \succ_i \delta_y$ for all $i \in I$, then $\delta_x \succ_0 \delta_y$.

Axiom 11.4 (Ex-Post Strong Pareto) if $\delta_x \succsim_i \delta_y$ for all $i \in I$ and $\delta_x \succ_i \delta_y$ for at least one $i \in I$, then $\delta_x \succ_0 \delta_y$.

The following result is straightforward since we only need to show that the society's vNM index is a monotone transformation of individual ones.

Theorem 11.2 *Suppose that $\succsim_i$ satisfies the expected utility theory for all $i \in I$ and $\succsim_0$ satisfies the theory too. For each $i \in I$ fix a function $u_i : X \to \mathbb{R}$ that forms an expected utility representation of $\succsim_i$, and fix $u_0 : X \to \mathbb{R}$ that forms an expected utility representation of $\succsim_0$. Assume the minimal agreement condition. Then:*

(i) $(\succsim_i)_{i \in \{0\} \cup I}$ satisfy Ex-post Weak Pareto if and only if there is a function $\varphi : \prod_{i \in I} u_i(X) \to \mathbb{R}$ such that $\varphi(u) > \varphi(v)$ holds whenever $u_i > v_i$ for all $i \in I$ and

$$u_0(x) = \varphi(u_1(x), \cdots, u_I(x))$$

for all $x \in X$.

(ii) $(\succsim_i)_{i \in \{0\} \cup I}$ satisfy Ex-post Strong Pareto if and only if there is a function $\varphi : \prod_{i \in I} u_i(X) \to \mathbb{R}$ such that $\varphi(u) > \varphi(v)$ holds whenever $u_i \geq v_i$ for all $i \in I$ and $u_i > v_i$ for at least one $i \in I$ and

$$u_0(x) = \varphi(u_1(x), \cdots, u_I(x))$$

for all $x \in X$.

11.1.3 Should Social Objective Satisfy the Expected Utility Theory?

We already saw that we should put a reservation on the meaning of welfare weights as obtained in the Harsanyi theorem and the meaning of having an additive form of aggregation.

We also need to question, before this, the requirement that the social ranking should satisfy the expected utility theory, in particular the independence axiom.

Let me revisit the example by Diamond [26].

Example 11.1 There are two prisoners, A and B. You have to execute one of the two tomorrow. You are indifferent between the two. Now suppose that you think you should respect fairness and think that flipping a fair coin is better than killing one of them with probability 1 arbitrarily.

This contradicts the expected utility theory. For, according to the independence axiom taking the even-chances mixture does not change the level of desirability when you are indifferent between killing A with probability 1 and killing B with probability 1.

As far as we assume that a social welfare function must satisfy the expected utility theory, there is no point in pursuing fairness of chances or equality of chances. In other words, if we want to pursue fairness of chances and equality of chances, we must deviate from the independence axiom.

Epstein and Segal [32] consider that the social ranking should satisfy an axiom weaker than Independence, called Mixture Symmetry, which was proposed for individual decision in Chew et al. [18].

Mixture Symmetry: For all $l, l' \in \Delta(X)$ and $\lambda \in [0, 1]$,

$$l \sim_0 l' \implies \lambda l + (1-\lambda)l' \sim_0 (1-\lambda)l + \lambda l'.$$

Theorem 11.3 *For each $i \in I$, assume that $\succsim_i$ satisfies the expected utility theory for all and fix its mixture-linear representation U_i. Then, the social ranking $\succsim_0$ satisfies Completeness, Transitivity, Mixture Continuity, Mixture Symmetry and $(\succsim_i)_{i \in I \cup \{0\}}$ satisfies Strong Pareto if and only if there exist a matrix $\mu \in \mathbb{R}^{I \times I}$ and a vector $\lambda \in \mathbb{R}^I$ such that $\succsim_0$ is represented in the form*

$$U_0(p) = W((U_i(p))_{i \in I}) = \sum_{i \in I} \sum_{j \in I} \mu_{ij} U_i(p) U_j(p) + \sum_{i \in I} \lambda_i U_i(p),$$

where $W : \{(U_i(p))_{i \in I} \in \mathbb{R}^I : p \in \Delta(X)\} \to \mathbb{R}$ is a strongly monotone function.

However, violating the independence axiom leads to violation of dynamic consistency. Let me cite again the example by Machina [76].

Example 11.2 ("Machina's Mom") From Machina (1989):

Mom has a single indivisible item — a "treat" — which she can give to either daughter Abigail or son Benjamin. Assume that she is indifferent between Abigail getting the treat and Benjamin getting the treat, and strongly prefers either of these outcomes to the case where neither child gets it. However, in a violation of the precepts of expected utility theory, Mom strictly prefers a coin flip over either of these sure outcomes, and in particular, strictly prefers 1/2: 1/2 to any other pair of probabilities. This random allocation procedure would be straightforward, except that Benjie, who cut his teeth on Raiffa's classic Decision Analysis, behaves as follows:

> Before the coin is flipped, he requests a confirmation from Mom that, yes, she does strictly prefer a 50:50 lottery over giving the treat to Abigail. He gets her to put this in writing. Had he won the flip, he would have claimed the treat. As it turns out, he loses the flip. But as Mom is about to give the treat to Abigail, he reminds Mom of her preference for flipping a coin over giving it to Abigail (producing her signed statement), and demands that she flip again.

> What would your Mom do if you tried to pull a stunt like this? She would undoubtedly say "You had your chance!" and refuse to flip the coin again. This is precisely what Mom does.

■

Machina continues, "By replying 'You had your chance,' Mom is reminding Benjamin of the existence of the snipped-off branch (the original 1/2 probability of B) and that her preferences are not separable, so the fact that nature could have gone down that branch still matters. Mom is rejecting the property of consequentialism — and, in my opinion, rightly so."

Each of Mom's and Benjamin's claims amounts to a problem. If we follow Mom's claim to reject consequentialism, the ex-post decision may depend on an event which did not happen and an outcome which would have happened in such an event. How can we model such a decision in an appealing way by itself, beyond simply mechanically taking the "projection" of the ex-ante optimal decision?

On the other hand, if we accept Benjamin's claim, as far as we want our welfare judgment to be dynamically consistent, it implies that the ex-ante judgment must support Abigail winning the item with probability $1/2 \times 1/2 = 1/4$ and Benjamin winning with probability $1/2 + 1/2 \times 1/2 = 3/4$, which is unfair in any sense from the ex-ante viewpoint.

Once we abandon consequentialism we can abandon the independence axiom without violating dynamic consistency. However, how can saying "You had your chance" be different from just saying "Sorry, we have already decided"? Do we follow the ex-ante plan merely because we have already decided? Is there any "thinking" here? Or "freedom," at least "freedom to reconsider?" Although it was not a successful attempt, Hayashi [59] might be a good starting point to think about this issue.

11.2 Social Decision Under Subjective Uncertainty*

Now we consider social decision making under subjective uncertainty, in which probability distributions over outcomes are not given as objects.

There are two kinds of disagreement here. One is about evaluation of outcomes, which has already occurred before. The other is about probabilistic assessment on states and events.

Under such double disagreements, the Pareto criterion is not obvious. Here is an example from Gilboa et al. [46].

Example 11.3 The problem is whether A and B should duel. A believes he will win and he is happy if he wins, hence he wants to duel. B believes he will win and he is happy if he wins, hence he wants to duel.

According to the Pareto principle, we should allow them to duel since both want to. But this is absurd because one of the must be eventually wrong.

This is an example of what Mongin [80] calls *spurious unanimity*.

11.2.1 Aggregation of Subjective Beliefs

First, let us ignore the disagreement about tastes over outcomes and focus on how to aggregate subjective probability distributions over events and states.

Let Ω denote the set of states and let Σ denote the family of events. Let I denote the set of individuals, and for each $i \in I$, let p_i denote his subjective probability distribution defined over (Ω, Σ). Also, let p_0 denote the society's probability distribution held over (Ω, Σ). Assume that $(p_1, \cdots, p_I)$ and p_0 are convex-ranged finitely additive probability measures.

Consider the following axioms.

Axiom 11.5 (Weak Pareto) if $p_i(A) > p_i(B)$ for all $i \in I$, then $p_0(A) > p_0(B)$.

Axiom 11.6 (Strong Pareto) if $p_i(A) \geq p_i(B)$ for all $i \in I$, then $p_0(A) \geq p_0(B)$, and if $p_i(A) > p_i(B)$ for at least one $i \in I$ additionally, then $p_0(A) > p_0(B)$.

The following mathematical result is known (Theorem 11.4.9, Rao and Rao [88]).

Proposition 11.1 (Lyapunov's Theorem) *If $\{p_i\}_{i \in I}$ is a collection of convex-ranged finitely additive probability measures over (Ω, Σ), then the set*

$$R((p_i)_{i \in I}) \equiv \{(p_1(E))_{i \in I} \in [0, 1]^I : E \in \Sigma\}$$

is convex.

Its usefulness is not immediate to us, but note that because $\mathbf{0}_I = (p_i(\emptyset))_{i \in I}$ and $\mathbf{1}_I = (p_i(\Omega))_{i \in I}$ both belong to $R((p_i)_{i \in I})$. From the above theorem this set is convex, for any $\eta \in [0, 1]$, $\eta \mathbf{1}_I \in R(R((p_i)_{i \in I}))$. Thus we obtain the following.

Lemma 11.1 *If $\{p_i\}_{i \in I}$ is a collection of convex-ranged finitely additive probability measures over (Ω, Σ), then for all $\eta \in [0, 1]$ there is $E \in \Sigma$ such that*

$$p_i(E) = \eta, \quad i \in I.$$

This is a surprising statement. For, if the state space is "rich" and each individual's subjective belief is represented by a convex-ranged finitely additive probability measure there always exists an event on which everybody assigns the same probability value, and it is true for any probability value. We denote the family of such "objective" events by

$$\Sigma^* = \{E \in \Sigma : p_i(E) = p_j(E), \ \forall i, j \in I\}.$$

The following is an immediate consequence.

Lemma 11.2 *If $\{p_i\}_{i \in I}$ is a collection of convex-ranged finitely additive probability measures over (Ω, Σ), then for all $\eta_1, \cdots, \eta_m \geq 0$ with $\sum_{k=1}^m \eta_k = 1$ there is a partition $E_1, \cdots, E_m \in \Sigma$ of Ω such that*

$$p_i(E_k) = \eta_k, \quad i \in I, \ k = \cdots, m.$$

Proposition 11.2 *A collection of convex-ranged finitely additive measures $p_0, \{p_i\}_{i \in I}$ defined over (Ω, Σ) satisfies*

$$p_i(A) > p_i(B) \ \forall i \in I \quad \Longrightarrow \quad p_0(A) > p_0(B)$$

for all $A, B \in \Sigma$ if and only if there is a vector $\alpha \in \mathbb{R}^I_+ \setminus \{\mathbf{0}\}$ with $\sum_{i \in I} \alpha_i = 1$ such that

$$p_0 = \sum_{i \in I} \alpha_i p_i$$

holds.

Proof Let

$$C = \{(p_0(E) - p_0(E'), (p_i(E) - p_i(E'))_{i \in I} \in \mathbb{R}^{\{0\} \cup I} : E, E' \in \Sigma\}$$

then from the Lyapunov theorem it is compact and convex. From the Pareto property,

$$C \cap \left((-\mathbb{R}_+) \times \mathbb{R}^I_{++}\right) = \emptyset.$$

Hence by the separating hyperplane theorem there is a vector $q \in \mathbb{R}^{\{0\} \cup I} \setminus \{\mathbf{0}_{\{0\} \cup I}\}$ and a number α such that

$$qx \geq \alpha \geq qy, \quad \forall x \in C, \ \forall y \in (-\mathbb{R}_+) \times \mathbb{R}^I_{++}.$$

Since $x \in C$ implies $-x \in C$, we have $qx \geq \alpha$ and $q(-x) \geq \alpha$, which implies $\alpha = 0$ and

$$qx = 0, \quad \forall x \in C.$$

Then the separation condition says $0 \geq qy$ for all $y \in (-\mathbb{R}_+) \times \mathbb{R}^I_{++}$. Suppose $q_i > 0$ for some $i \in I$. Then, by taking $y_i = 1$, by letting $y_0 = -\varepsilon$ and $y_j = \varepsilon$ for all $j \in I \setminus \{i\}$ for sufficiently small $\varepsilon > 0$, we have

$$qy = q_i - \varepsilon \left(q_0 + \sum_{j \in I \setminus \{i\}} q_j \right) > 0,$$

which contradicts the separation condition. Thus $q_i \leq 0$ holds for all $i \in I$.

Also, suppose $q_0 < 0$. Then, by taking $y_0 = -1$ and $y_i = \varepsilon$ for all $i \in I$ for sufficiently small $\varepsilon > 0$, we have

$$qy = -q_0 + \varepsilon \sum_{i \in I} q_i > 0,$$

which contradicts to the separation condition. Hence $q_0 \geq 0$.

Now suppose $q_0 = 0$. Then, since $q \in \mathbb{R}^{\{0\} \cup I} \setminus \{\mathbf{0}_{\{0\} \cup I}\}$ we must have $q_I \in -\mathbb{R}^I_+ \setminus \{\mathbf{0}_I\}$. Because $qx = 0$ holds for all $x \in C$,

$$0 = \sum_{i \in I} q_i (p_i(E) - p_i(E'))$$

holds for all $E, E' \in \Sigma$. Thus we obtain a contradiction by taking $E = \Omega$ and $E' = \emptyset$. Hence $q_0 > 0$.

Suppose $q_i = 0$ for all $i \in I$. Then we have $q_0(p_0(E) - p_0(E'))$ for all $E, E' \in \Sigma$. Then we obtain a contradiction by taking $E = \Omega$ and $E' = \emptyset$. Hence $q_i < 0$ for at least one $i \in I$.

From the above, there exist $q_0 > 0$ and $q_I \in -\mathbb{R}^I_+ \setminus \{\mathbf{0}_I\}$ such that

$$q_0(p_0(E) - p_0(E')) = \sum_{i \in I} -q_i (p_i(E) - p_i(E'))$$

holds for all $E, E' \in \Sigma$.

Now $\lambda_i = -q_i/q_0$ for each $i \in I$. Fix some $E' \in \Sigma$ and let $\mu_0 = p_0(E') + \sum_{i \in I} -(q_i/q_0) p_i(E')$. Then

$$p_0 = \sum_{i \in I} \lambda_i p_i + \mu_0.$$

Finally, because p_0 is a probability measure it follows $\mu_0 = 0$ and $\sum_{i \in I} \lambda_i = 1$. □

11.2.2 Aggregation of Subjective Expected Utility Preferences

Now consider the problem aggregation of preferences over bets, Savage acts. We continue to denote pair of set of states and family of events by (Ω, Σ). We denote the set of social outcomes by C. Then a simple Savage act is a mapping from Ω to X, denoted typically by $f : \Omega \to X$, such that $f(\Omega)$ is finite and for all $x \in f(\Omega)$ it holds $f^{-1}(x) \in \Sigma$. Denote the set of simple Savage acts by $\mathcal{F}$.

For each $i \in I$, denote his preference over simple Savage acts by $\succsim_i$, and denote the social ranking over simple Savage acts by $\succsim_0$.

For each $i \in I$, we assume that his preference $\succsim_i$ follows the subjective expected utility (SEU) theory and it is represented in the form

$$U_i(f) = \sum_{x \in f(\Omega)} u_i(x) p_i(f^{-1}(x)),$$

where U_i denotes the entire SEU representation, u_i is the corresponding vNM index and p_i is his subjective belief as a convex-ranged finitely additive measure. Denote such a triple by (U_i, u_i, p_i). Note, the assumption that each individual's preference follows the SEU theory is a *descriptive* one.

We also assume that the social ranking $\succsim_0$ follows the SEU theory and it is represented in the form

$$U_0(f) = \sum_{x \in f(\Omega)} u_0(x) p_0(f^{-1}(x)),$$

where U_0 denotes the entire SEU representation, u_0 is the corresponding vNM index of the society and p_0 is the social belief as a convex-ranged finitely additive probability measure. Denote such a triple by (U_0, u_0, p_0). Note, the assumption that the social ranking follows the SEU theory is a *normative requirement*.

Axiom 11.7 (Ex-Ante Weak Pareto) For all $f, g \in \mathcal{F}$, if $f \succ_i g$ for all $i \in I$, then $f \succ_0 g$.

Axiom 11.8 (Ex-Ante Strong Pareto) For all $f, g \in \mathcal{F}$, if $f \succsim_i g$ for all $i \in I$ and $f \succ_i g$ for some $i \in I$, then $f \succ_0 g$.

Throughout, we assume the following minimal agreement condition, which is natural in the context of resource allocation.

Minimal Agreement Condition (MAC): There exist c, c' such that $c \succ_i c'$ holds for all $i \in I$.

Theorem 11.4 *Assume that $\succsim_i$ follows the SEU theory for all $i \in I$ and $\succsim_0$ follows the SEU theory as well. For each $i \in I$, fix a triple (U_i, u_i, p_i) that gives an SEU representation of $\succsim_i$. Also, fix a triple (U_0, u_0, p_0) that gives an SEU representation of $\succsim_0$. Then:*

(i) $(\succsim_i)_{i\in\{0\}\cup I}$ *satisfies Ex-ante Weak Pareto if and only if there exist a vector* $\lambda \in \mathbb{R}^I_+ \setminus \{\mathbf{0}\}$ *and a number* μ_0 *and a vector* $\alpha \in \mathbb{R}^I_+ \setminus \{\mathbf{0}\}$ *satisfying* $\sum_{i\in I} \alpha_i = 1$ *such that*

$$U_0 = \sum_{i\in I} \lambda_i U_i + \mu_0,$$

$$u_0 = \sum_{i\in I} \lambda_i u_i + \mu_0,$$

$$p_0 = \sum_{i\in I} \alpha_i p_i$$

holds.

(ii) However, if one of the following holds there exists $i \in I$ *such that* $\lambda_i > 0$, $\alpha_i = 1$ *and* $\lambda_j = 0$, $\alpha_j = 0$ *for all* $j \neq i$.

1. *For every* i*, there exist* $x, y \in X$ *such that* $u_i(x) > u_i(y)$ *and* $u_j(y) > u_j(x)$ *for all* $j \neq i$.
   Also, for all i*, there exists* E *such that* $P_i(E) > P_j(E) = P_{j'}(E)$ *for all* $j, j' \neq i$.
2. *For every* i*, there exist* $x, y \in X$ *such that* $u_i(x) > u_i(y)$ *and* $u_j(y) = u_j(x)$ *for all* $j \neq i$.
   Also, for all i*, there exists* E *such that* $P_i(E) > P_j(E)$ *for all* $j \neq i$.

(iii) Under (ii), there is no (U_0, u_0, p_0) *satisfying Ex-ante Strong Pareto.*

Proof (i) We show that the set

$$C = \{(U_0(f) - U_0(f'), (U_i(f) - U_i(f'))_{i\in I}) \in \mathbb{R}^{\{0\}\cup I} : f, f' \in \mathcal{F}\}$$

is convex, and then the argument similar to the one for Harsanyi's theorem follows.

Denote the partition of the state space given by f by $A_1, \cdots, A_m$, the one given by $\widetilde{f}$ by $B_1, \cdots, B_n$, the one given by g by $C_1, \cdots, C_r$, and the one given by $\widetilde{g}$ by $D_1, \cdots, D_s$.

Then, by applying the Lyapunov's theorem to $(p_0, (p_i)_{i\in I})$, we can take $S, \widetilde{S} \subset \Omega$ such that

$$p_i(A_k \cap S) = \eta p_i(A_k), \quad p_i(C_l \cap S^c) = (1-\eta) p_i(C_l),$$
$$k = 1, \cdots, m, \ l = 1, \cdots, r,$$

$$p_i(B_k \cap \widetilde{S}) = \eta p_i(B_k), \quad p_i(D_l \cap \widetilde{S}^c) = (1-\eta) p_i(D_l),$$
$$k = 1, \cdots, n, \ l = 1, \cdots, s,$$

where $i \in \{0\} \cup I$.

Now consider h, h' given by

$$h(\omega) = \begin{pmatrix} f(\omega) & \omega \in S \\ g(\omega) & \omega \in S^c \end{pmatrix}, \quad \widetilde{h}(\omega) = \begin{pmatrix} \widetilde{f}(\omega) & \omega \in \widetilde{S} \\ \widetilde{g}(\omega) & \omega \in \widetilde{S}^c \end{pmatrix}.$$

Then

$$U_i(h) = \eta U_i(f) + (1 - \eta)U_i(g),$$

$$U_i(\widetilde{h}) = \eta U_i(\widetilde{f}) + (1 - \eta)U_i(\widetilde{g}),$$

where $i \in \{0\} \cup I$, which establishes convexity of set C.

Thus, under Ex-ante Weak Pareto, the Harsanyi theorem argument establishes existence of a vector $\lambda \in \mathbb{R}^I_+ \setminus \{\mathbf{0}_I\}$ and a number μ_0, where

$$U_0 = \sum_{i \in I} \lambda_i U_i + \mu_0.$$

By restricting this to X, we obtain

$$u_0 = \sum_{i \in I} \lambda_i u_i + \mu_0.$$

By the minimal agreement condition, fix $x, y \in X$ such that $x \succ_i y$ holds for all $i \in I$. Then we can see that

$$p_i(A) \geq p_i(B) \quad \Longleftrightarrow \quad \begin{pmatrix} x & A \\ y & A^c \end{pmatrix} \succsim_i \begin{pmatrix} x & B \\ y & B^c \end{pmatrix}$$

follows from the subjective expected utility theory. Hence Ex-ante Weak Pareto implies the weak Pareto property in the probability aggregation problem. Similarly for the case of the ex-ante strong Pareto property.

Hence from Proposition 11.2, p_0 is written as a convex combination of $\{p_i\}_{i \in I}$.
(ii) Consider a bet

$$f = \begin{pmatrix} x & E \\ y & E^c \end{pmatrix},$$

then it follows that

$$\begin{aligned} U_0(f) &= \sum_{i \in I} \lambda_i U_i(f) \\ &= \sum_{i \in I} \lambda_i \left\{ u_i(x) p_i(E) + u_i(y) p_i(E^c) \right\} \\ &= \sum_{i \in I} \lambda_i \left\{ u_i(x) - u_i(y) \right\} p_i(E) + \sum_{i \in I} \lambda_i u_i(y) \end{aligned}$$

On the other hand, since

$$\begin{aligned} U_0(f) &= u_0(x) p_0(E) + u_0(y) p_0(E^c) \\ &= \left\{ u_0(x) - u_0(y) \right\} p_0(E) + u_0(y) \\ &= \left[\sum_{i \in I} \lambda_i \left\{ u_i(x) - u_i(y) \right\} \right] \left[\sum_{i \in I} \alpha_i p_i(E) \right] + \sum_{i \in I} \lambda_i u_i(y), \end{aligned}$$

by combining with the above we obtain

$$\sum_{i \in I} \lambda_i \left\{ u_i(x) - u_i(y) \right\} p_i(E) = \left[\sum_{i \in I} \lambda_i \left\{ u_i(x) - u_i(y) \right\} \right] \left[\sum_{i \in I} \alpha_i p_i(E) \right].$$

By combining this with $\sum_{i=1}^{n} \alpha_i = 1$ we have

$$\sum_{i \in I} \sum_{j \neq i} \lambda_i \alpha_j \left\{ u_i(x) - u_i(y) \right\} \left\{ p_i(E) - p_j(E) \right\} = 0.$$

By applying Condition 1, pick any $i \in I$ and take $x, y \in X$ such that $u_i(x) - u_i(y) > 0$ and $u_j(x) - u_j(y) < 0$ for all $j \neq i$, take E such that $p_i(E) > p_j(E) = p_k(E)$ for all $j, k \neq i$, then

$$\lambda_i \alpha_j = 0$$

holds for all $j \neq i$.

Since i was arbitrary, for all i and for all $j \neq i$ we have

$$\lambda_i \alpha_j = 0.$$

Now suppose $\lambda_i > 0$, then for all $j \neq i$ we have $\alpha_j = 0$, hence $\alpha_i = 1$. Thus it is impossible that $\lambda_i, \lambda_{i'} > 0$ for different i, i', meaning that there is only one i such that $\lambda_i > 0$, and we have $\alpha_i = 1$ for such an i.

Similarly for the case that we apply Condition 2.

(iii): Clear from (ii). □

11.3 The Ex-Ante Pareto Criterion and State-Dependence of Social Decision

Where exactly is the conflict in the above impossibility result at a conceptual level? Or, which axiom of the subjective expected utility theory is violated by Pareto-respecting social decision criteria?

Chambers and Hayashi [15] showed that it is P3 and P4 of the Savage axioms. L

Example 11.4 A believes that it will be sunny tomorrow for sure. B believes it will rain tomorrow for sure. According to the ex-ante Pareto criterion, A should get all the resources if it is sunny tomorrow and B should get all if it rains tomorrow. Then, it means that ex-post welfare weight is 100% for A if it is sunny and 100% for B if it rains, and this violates state-independence of outcome evaluation, Savage's P3.

Think of the following example as well.

Example 11.5 A believes that event D is more likely than D^c, and B believes the opposite. For simplicity, assume that outcomes are monetary amounts, where each individual prefers to receive more. Let (x_A, x_B) denote an allocation that A receives x_A and B receives x_B. Then by symmetry it is natural to say

$$(100, 0) \sim_0 (0, 100).$$

Now, since

$$\begin{pmatrix} (100, 0) & D \\ (0, 100) & D^c \end{pmatrix} \succ_A \begin{pmatrix} (0, 100) & D \\ (100, 0) & D^c \end{pmatrix},$$

$$\begin{pmatrix} (100, 0) & D \\ (0, 100) & D^c \end{pmatrix} \succ_B \begin{pmatrix} (0, 100) & D \\ (100, 0) & D^c \end{pmatrix},$$

from Ex-ante Weak Pareto,

$$\begin{pmatrix} (100, 0) & D \\ (0, 100) & D^c \end{pmatrix} \succ_0 \begin{pmatrix} (0, 100) & D \\ (100, 0) & D^c \end{pmatrix}.$$

However, from $(100, 0) \sim_0 (0, 100)$ and P3 we deduce

$$\begin{pmatrix} (100, 0) & D \\ (0, 100) & D^c \end{pmatrix} \sim_0 (0, 100)$$

and

$$\begin{pmatrix} (0, 100) & D \\ (100, 0) & D^c \end{pmatrix} \sim_0 (100, 0),$$

which imply by transitivity that

$$\begin{pmatrix} (100,0) & D \\ (0,100) & D^c \end{pmatrix} \sim_0 \begin{pmatrix} (0,100) & D \\ (100,0) & D^c \end{pmatrix},$$

which is a contradiction.

Generalization of the above example leads to the following result.

Theorem 11.5 *Suppose that* $\succsim_A$ *and* $\succsim_B$ *satisfy P4, and that for each* $i = A, B$ *and events* D, E,

$$D \succsim_i^l E \iff D^c \precsim_i^l E^c.$$

Also, assume that there exist $\overline{x}, \underline{x} \in X$ *such that* $\overline{x} \succ_A \underline{x}$, $\overline{x} \prec_B \underline{x}$ *and* $\overline{x} \sim_0 \underline{x}$.

Then, the social ranking $\succsim_0$ *satisfies P3 and Ex-ante Strong Pareto only if* $\succsim_A^l = \succsim_B^l$.

Proof Suppose there exist D, E such that $D \succsim_A^l E$ and $D \prec_B^l E$.

Then, since each individual's preference satisfies P4 and $D^c \succ_B^l E^c$ follows from the assumption, we have $\overline{x}D\underline{x} \succsim_A \overline{x}E\underline{x}$, $\overline{x}D\underline{x} \succ_B \overline{x}E\underline{x}$. Hence by Ex-ante Strong Pareto it follows that $\overline{x}D\underline{x} \succ_0 \overline{x}E\underline{x}$.

On the other hand, since $\overline{x} \sim_0 \underline{x}$, from P3 for $\succsim_0$ we have $\overline{x}D\underline{x} \sim_0 \underline{x}$ and $\overline{x}E\underline{x} \sim_0 \underline{x}$, which implies $\overline{x}D\underline{x} \sim_0 \overline{x}E\underline{x}$ by transitivity, a contradiction. □

Example 11.6 Again, A believes that event D is more likely than D^c, and B believes the opposite. For simplicity, assume that outcomes are monetary amounts, where each individual prefers to receive more. Let (x_A, x_B) denote an allocation that A receives x_A and B receives x_B.

Then, since

$$\begin{pmatrix} (100,0) & D \\ (0,0) & D^c \end{pmatrix} \succ_A \begin{pmatrix} (100,0) & E \\ (0,0) & E^c \end{pmatrix},$$

$$\begin{pmatrix} (100,0) & D \\ (0,0) & D^c \end{pmatrix} \sim_B \begin{pmatrix} (100,0) & E \\ (0,0) & E^c \end{pmatrix},$$

it follows from Ex-ante Strong Pareto that

$$\begin{pmatrix} (100,0) & D \\ (0,0) & D^c \end{pmatrix} \succ_0 \begin{pmatrix} (100,0) & E \\ (0,0) & E^c \end{pmatrix},$$

which reveals that the society believes D is more likely than E according to P4.

On the other hand, since

$$\begin{pmatrix} (0,100) & D \\ (0,0) & D^c \end{pmatrix} \sim_A \begin{pmatrix} (0,100) & E \\ (0,0) & E^c \end{pmatrix},$$

$$\begin{pmatrix} (0,100) & D \\ (0,0) & D^c \end{pmatrix} \prec_B \begin{pmatrix} (0,100) & E \\ (0,0) & E^c \end{pmatrix},$$

it follows from Ex-ante Strong Pareto that

$$\begin{pmatrix} (0,100) & D \\ (0,0) & D^c \end{pmatrix} \prec_0 \begin{pmatrix} (0,100) & E \\ (0,0) & E^c \end{pmatrix},$$

which reveals that the society believes E is more likely than D according to P4. Thus we obtain a contradiction.

Generalization of the above example leads to the following result.

Theorem 11.6 *Assume that* $\succsim_A$, $\succsim_B$ *satisfy P4, and that for each* $i = A, B$ *and for all* D, E *and* x, x',

$$x \sim_i x' \implies xDx' \sim_i xEx'.$$

Also, assume that there exist $\overline{x}, \underline{x}, \overline{y}, \underline{y} \in X$ *such that* $\overline{x} \succ_A \underline{x}$, $\overline{x} \sim_A \underline{x}$, $\overline{y} \sim_B \underline{y}$, $\overline{y} \succ_B \underline{y}$.

Then, $\succsim_0$ *satisfies P4 and Ex-ante Strong Pareto only if* $\succsim_A^l = \succsim_B^l$.

Proof Suppose there exist events D, E such that $D \succsim_A^l E$ and $D \prec_B^l E$. Then by the assumption, $\overline{x}D\underline{x} \succsim_A \overline{x}E\underline{x}$, $\overline{x}D\underline{x} \sim_B \overline{x}E\underline{x}$. Hence by Ex-ante Strong Pareto we have $\overline{x}D\underline{x} \succsim_0 \overline{x}E\underline{x}$, revealing $D \succsim_0^l E$.

On the other hand, again by the assumption, $\overline{y}D\underline{y} \sim_A \overline{y}E\underline{y}$, $\overline{y}D\underline{y} \prec_B \overline{y}E\underline{y}$. Hence by Ex-ante Strong Pareto again we have $\overline{y}D\underline{y} \prec_0 \overline{y}E\underline{y}$, revealing $D \prec_0^l E$, which is a contradiction. □

11.4 Weakening Pareto

Since it is impossible for a social decision criterion to satisfy the subjective expected utility theory (especially state-independence of outcome evaluation) and the ex-ante Pareto condition without avoiding dictatorship, we have to give up either one.

If we allow state-dependence of outcome evaluation, it is shown by Chambers and Hayashi [15] that additive aggregation follows from the ex-ante Pareto criterion. Is there anything wrong with allowing state dependence at a social level? To see this, consider additive aggregation

$$U_0(f) = \sum_{i \in I} \alpha_i U_i(f) = \sum_{i \in I} \alpha_i \sum_{k=1}^{m} u_i(x_k) p_i(E_k),$$

where $f = (x_1; E_1, \cdots, x_m; E_m)$, which can be rewritten as

$$U_0(f) = \sum_{k=1}^{m} \left\{ \sum_{i \in I} \alpha_i(E_k) u_i(x_k) \right\} p_0(E_k),$$

where

$$p_0(E) = \sum_{i \in I} \alpha_i p_i(E)$$

and

$$\alpha_i(E) = \frac{\alpha_i p_i(E)}{\sum_{j \in I} \alpha_j p_j(E)}.$$

This means that the decision power of each individual at each event ex-post is proportional to the prior probability she assigned to that event. Therefore, if she has failed to predict some event in the sense that she assigned low probability to it, she has no say once such event has happened. Maybe there is nothing wrong in the market, but it is not the case in voting, in which one person has one vote no matter whether he or she was successful in predicting events in the past. At least, we should expect political tension in enforcing an ex-ante socially "optimal" decision in the ex-post stage.

How about weakening Pareto? Chambers and Hayashi [16] considered a weaker version of the Pareto condition in an incomplete-information setting, stating "if it is commonly known that f is better than g for all people the society should rank f over g," but show that it leads to the same impossibility. This means that the impossibility result is quite strong.

Not just because of the impossibility result, the ex-ante Pareto axiom may be seen as unreasonable, because unanimity may come from double disagreements on beliefs and tastes as discussed above.

11.4.1 The Pareto Criterion Applied when There Is No Disagreement on Beliefs

Here we follow Gilboa et al. [46] and consider that the Pareto condition should apply only when everybody agrees on the likelihood of relevant events.

Are there such nice situations, generically? Yes, when the set of states of the world is rich and each individual's subjective belief takes the form of convex-ranged finitely probability measures. Following the Lyapunov theorem, there is a rich class of such events. Denote such a class by

$$\Sigma^* = \{E \in \Sigma : p_i(E) = p_j(E), \ \forall i, j \in I\}$$

and let

$$\mathcal{F}^* = \{f \in \mathcal{F} : f^{-1}(B) \in \Sigma^*, \ \forall B \subset X, \ \ |B| < \infty\}$$

denote the set of acts that are measurable with respect to such events.

We impose the following Pareto condition on such acts.

Axiom 11.9 (Consensus Weak Pareto) For all $f, g \in \mathcal{F}^*$, if $f \succ_i g$ for all $i \in I$, then $f \succ_0 g$.

Axiom 11.10 (Consensus Strong Pareto) Fo all $f, g \in \mathcal{F}^*$, if $f \succsim_i g$ for all $i \in I$, then $f \succsim_0 g$. If $f \succ_i g$ additionally for at least one $i \in I$, then $f \succ_0 g$.

Theorem 11.7 *Assume that $\succsim_i$ satisfies the subjective expected utility theory for $i \in I$ and $\succsim_0$ as well. For each $i \in I$, fix a pair (u_i, p_i) giving an SEU representation of $\succsim_i$ and fix a pair (u_0, p_0) giving a SEU representation of $\succsim_0$, where $(p_i)_{i \in \{0\} \cup I}$ are finitely additive and convex-ranged. Assume the Minimal Agreement Condition.*

Then, $(\succsim_i)_{i \in \{0\} \cup I}$ satisfy Consensus Weak Pareto if and only if there exist a vector $\lambda \in \mathbb{R}^I_+ \setminus \{\mathbf{0}_I\}$ and a number μ_0 and a vector $\alpha \in \mathbb{R}^I_+ \setminus \{\mathbf{0}_I\}$ with $\sum_{i \in I} \alpha_i = 1$ such that

$$u_0 = \sum_{i \in I} \lambda_i u_i + \mu_0,$$

$$p_0 = \sum_{i \in I} \alpha_i p_i$$

holds.

The same claim holds with $\lambda \in \mathbb{R}^I_{++}$ under Consensus Strong Pareto.

Proof Necessity of the Pareto condition is straightforward, so we show sufficiency.

The fact that u_0 is given by a weighted sum of $u_1, \cdots, u_I$ reduces to aggregation of preferences over lotteries induced by acts in $\mathcal{F}^*$, hence the previous theorem applies.

To show that p_0 is given by a convex combination of $\{p_i\}_{i \in I}$, consider sets

$$Z = \{(p_i(E))_{i \in I} \in [0, 1]^I : E \in \Sigma\},$$
$$Z^* = \{(p_0(E), (p_i(E))_{i \in I}) \in [0, 1]^{\{0\} \cup I} : E \in \Sigma\}.$$

By the Lyapunov theorem, Z and Z^* are compact and convex.

Claim If $\left(w_0, \frac{1}{2}\mathbf{1}_I\right) \in Z^*$, then $w_0 = \frac{1}{2}$.

Suppose $w_0 \neq \frac{1}{2}$. Without loss, suppose $w_0 > \frac{1}{2}$. Then, by convexity of Z^*,

$$(1-\lambda)\left(w_0, \frac{1}{2}\mathbf{1}_I\right) = \lambda\mathbf{0}_{I+1} + (1-\lambda)\left(w_0, \frac{1}{2}\mathbf{1}_I\right) \in Z^*,$$

and for sufficiently small $\lambda > 0$ we have $(1-\lambda)w_0 > \frac{1}{2}$ and $(1-\lambda)\frac{1}{2} < \frac{1}{2}$.

Now, let E be an event such that $(p_0(E), (p_i(E))_{i\in I}) = (1-\lambda)\left(w_0, \frac{1}{2}\mathbf{1}_I\right)$, then $p_i(E) < \frac{1}{2}$ holds for all $i \in I$ and $p_0(E) > \frac{1}{2}$.

Let $x, y \in X$ be such that $x \succ_i y$ for all $i \in I$. Then

$$\begin{pmatrix} x & E \\ y & E^c \end{pmatrix} \prec_i \begin{pmatrix} x & E^c \\ y & E \end{pmatrix}$$

holds for all $i \in I$, but we have

$$\begin{pmatrix} x & E \\ y & E^c \end{pmatrix} \succ_0 \begin{pmatrix} x & E^c \\ y & E \end{pmatrix},$$

which is a contradiction.

Claim For all $z \in Z$, there is unique $w_0(z)$ such that $(w_0(z), z) \in Z^*$.

Suppose there are $w_0 > w_0'$ such that $(w_0, z), (w_0', z) \in Z^*$.

Then, $\mathbf{1}_{I+1} - (w_0', z)$ is nothing but the vector of probability assessments on the complement of the event giving (w_0', z), hence it belongs to Z^*. By convexity of Z^*, we have

$$\frac{1}{2}(w_0, z) + \frac{1}{2}\{\mathbf{1}_{\{0\}\cup I} - (w_0', z)\} = \left(\frac{1}{2} + \frac{1}{2}(w_0 - w_0'), \frac{1}{2}\mathbf{1}_I\right) \in Z^*,$$

which is a contradiction to Claim 11.4.1. □

Claim $w_0(\lambda z + (1-\lambda)z') = \lambda w_0(z) + (1-\lambda)w_0(z')$.

From convexity of Z^*,

$$\lambda(w_0(z), z) + (1-\lambda)(w_0(z'), z') \in Z^*,$$

which means

$$(\lambda w_0(z) + (1-\lambda)w_0(z'), \lambda z + (1-\lambda)z') \in Z^*.$$

Hence the conclusion follows from Claim 11.4.1.

Thus the function $w_0 : Z \to [0, 1]$ is mixture-linear and hence it takes the form

$$w_0(z) = \sum_{i \in I} \alpha_i z_i.$$

Since $w_0(\mathbf{1}_I) = 1$, we have $\sum_{i \in I} \alpha_i = 1$.

11.4.2 Hypothetical Exchange of Subjective Beliefs

The above possibility result depends critically on a rich state space, which is basically assumed to be a continuum. Hence it does not help for example when there is disagreement on subjective beliefs over a finite set of states.

Below, assume that the set of states Ω is finite. Let Z denote the set of pure outcomes and $\Delta_S(Z)$ denote the set of simple lotteries. Thus we consider Anscombe-Aumann acts (AA acts) in the form of a mapping $h : \Omega \to \Delta_S(Z)$. Let $\mathcal{H}$ denote the set of AA acts.

Let I denote the set of individuals, and for each $i \in I$ has preference defined over $\mathcal{H}$ follows the Anscombe-Aumann subjective expected utility theory and it is represented in the form

$$\sum_{\omega \in \Omega} u_i(f(\omega)) p_i(\omega),$$

where $u_i : \Delta_S(Z) \to \mathbb{R}$ is i's vNM index and $p_i \in \Delta(\Omega)$ is i's subjective belief.

We assume the richness assumption throughout.

Minimal Agreement Condition: There exist $l, l' \in \Delta_S(Z)$ such that $u_i(l) > u_i(l')$ for all $i \in I$.

The social ranking defined over $\mathcal{H}$ also follows the Anscombe-Aumann subjective expected utility theory and it is represented in the form

$$\sum_{\omega \in \Omega} u_0(f(\omega)) p_0(\omega),$$

where $u_0 : \Delta_S(Z) \to \mathbb{R}$ is the society's vNM index and $p_0 \in \Delta(\Omega)$ is the society's belief.

Below, given a mixture-linear function $u : \Delta_S(Z) \to \mathbb{R}$ and a probability distribution $p \in \Delta(\Omega)$, we denote the expected utility evaluation of an AA act $f \in \mathcal{H}$ by

$$\langle u \circ f, p \rangle = \sum_{\omega \in \Omega} u(f(\omega)) p(\omega).$$

Gayer et al. [39] and also Billot and Qu [12] introduced the following axiom.

Axiom 11.11 (Ex-Ante Pareto Under Belief Exchange) For all $f, g \in \mathcal{F}$, if $\langle u_i \circ f, p_j \rangle > \langle u_i \circ g, p_j \rangle$ for all $i, j \in I$, then $\langle u_0 \circ f, p_0 \rangle > \langle u_0 \circ g, p_0 \rangle$.

The axiom proposes that we calculate each individual's expected utility not just by using his own subjective belief but also all the other individuals' subjective beliefs. Hence the presumption of the axiom consists of $|I|^n$ conditions instead of $|I|$, meaning that the axiom is weaker than the traditional ex-ante Pareto condition.

The following possibility result is obtained under such weakening of the Pareto condition.

Theorem 11.8 *Assume that* $\succsim_i$ *satisfies the subjective expected utility theory for* $i \in I$ *and* $\succsim_0$ *as well. For each* $i \in I$*, fix a pair* (u_i, p_i) *giving an SEU representation of* $\succsim_i$ *and fix a pair* (u_0, p_0) *giving a SEU representation of* $\succsim_0$*. Assume the Minimal Agreement Condition. Then, Ex-ante Pareto under Belief Exchange is met if and only if there exist a vector* $\lambda \in \mathbb{R}^I_+ \setminus \{\mathbf{0}_I\}$*, a number* μ_0 *and a vector* $\alpha \in \mathbb{R}^I_+ \setminus \{\mathbf{0}_I\}$ *with* $\sum_{i \in I} \alpha_i = 1$ *such that*

$$u_0 = \sum_{i \in I} \lambda_i u_i + \mu_0,$$

$$p_0 = \sum_{i \in I} \alpha_i p_i$$

hold.

Proof Since the ex-ante Pareto condition holds over the set of lotteries $\Delta_S(Z)$, from the Harsanyi theorem there exist a vector $\lambda \in \mathbb{R}^I_+ \setminus \{\mathbf{0}_I\}$ and a number μ_0 such that

$$u_0 = \sum_{i \in I} \lambda_i u_i + \mu_0$$

holds.

To show the second part, suppose not and assume

$$p_0 \notin \left\{ \sum_{i \in I} \alpha_i p_i : \alpha \in \mathbb{R}^I_+ \setminus \{\mathbf{0}_I\}, \sum_{i \in I} \alpha_i = 1 \right\} \equiv P.$$

Then, since the right-hand-side is a closed convex set, by the separating hyperplane theorem there is a vector $v \in \mathbb{R}^\Omega \setminus \{\mathbf{0}_\Omega\}$ and a number β such that

$$v \cdot p_0 \geq \beta > v \cdot p \quad \forall p \in P$$

holds.

Let $\mathbf{1}_\Omega$be the $|\Omega|$-dimensional vector of ones, then

$$v \cdot p_0 \geq \beta \mathbf{1}_\Omega \cdot p_0 = \beta \mathbf{1}_\Omega \cdot p > v \cdot p \quad \forall p \in P.$$

Without loss of generality, take $f \in \mathcal{H}$ such that $v = u_i \circ f$ holds for all $i \in \{0\} \cup I$, and take $g \in \mathcal{H}$ such that $\beta \mathbf{1} = u_i \circ g$ holds for all $i \in \{0\} \cup I$. Then, $\langle u_i \circ f, p_j \rangle < \langle u_i \circ g, p_j \rangle$ holds for all $i, j \in I$, but we have $\langle u_0 \circ f, p_0 \rangle \geq \langle u_0 \circ g, p_0 \rangle$, which is a violation of Ex-ante Pareto under Belief Exchange. □

11.4.3 Pareto Under Certainty, Pareto Under Common Tastes

An alternative way is to impose the Pareto condition only on outcomes with certainty and on a subdomain of acts in which there is no disagreement in tastes (Danan et al. [19]).

Axiom 11.12 (Certainty Pareto) For all $l, m \in \Delta_S(Z)$, if $l \succ_i m$ for all $i \in I$, then $l \succ_0 m$.

Under the Minimal Agreement Condition, there exist $\underline{l}, \bar{l} \in \Delta_S(Z)$ such that $\bar{l} \succ_i \underline{l}$ for all $i \in I$. Let $\Delta(\underline{l}, \bar{l}) = \{\mu \bar{l} + (1-\mu)\underline{l} : \mu \in [0, 1]\}$, then

$$l \succsim_i m \iff l \succsim_j m$$

holds for all $i, j \in I$ and $l, m \in \Delta(\underline{l}, \bar{l})$. Now let $\mathcal{H}(\underline{l}, \bar{l}) = \Delta(\underline{l}, \bar{l})^S$ be the set of acts which take values only in $\Delta(\underline{l}, \bar{l})$. Thus any difference in preferences over $\mathcal{H}(\underline{l}, \bar{l})$ is attributed to difference in beliefs.

Axiom 11.13 (Common Taste Pareto) For all $f, g \in \mathcal{H}(\underline{l}, \bar{l})$, if $f \succ_i g$ for all $i \in I$, then $f \succ_0 g$.

Theorem 11.9 *Assume that $\succsim_i$ satisfies the subjective expected utility theory for $i \in I$ and $\succsim_0$ as well. For each $i \in I$, fix a pair (u_i, p_i) giving an SEU representation of $\succsim_i$ and fix a pair (u_0, p_0) giving a SEU representation of $\succsim_0$. Assume the Minimal Agreement Condition. Then, Certainty Pareto and Common Taste Pareto are met if and only if there exist a vector $\lambda \in \mathbb{R}^I_+ \setminus \{\mathbf{0}_I\}$, a number μ_0 and a vector $\alpha \in \mathbb{R}^I_+ \setminus \{\mathbf{0}_I\}$ with $\sum_{i \in I} \alpha_i = 1$ such that*

$$u_0 = \sum_{i \in I} \lambda_i u_i + \mu_0,$$

$$p_0 = \sum_{i \in I} \alpha_i p_i$$

hold.

Proof Since the Pareto condition holds over the set of lotteries $\Delta_S(Z)$, from the Harsanyi theorem there exist a vector $\lambda \in \mathbb{R}^I_+ \setminus \{\mathbf{0}_I\}$ and a number μ_0 such that

$$u_0 = \sum_{i \in I} \lambda_i u_i + \mu_0$$

holds.

To show the second part, suppose not and assume

$$p_0 \notin \left\{ \sum_{i \in I} \alpha_i p_i : \alpha \in \mathbb{R}^I_+ \setminus \{\mathbf{0}_I\}, \sum_{i \in I} \alpha_i = 1 \right\} \equiv P.$$

Then, since the right-hand side is a closed convex set, by the separating hyperplane theorem there is a vector $v \in \mathbb{R}^\Omega \setminus \{\mathbf{0}_\Omega\}$ and a number β such that

$$v \cdot p_0 \geq \beta > v \cdot p \quad \forall p \in P$$

holds.

Let $\mathbf{1}_\Omega$be the $|\Omega|$-dimensional vector of ones, then

$$v \cdot p_0 \geq \beta \mathbf{1}_\Omega \cdot p_0 = \beta \mathbf{1}_\Omega \cdot p > v \cdot p \quad \forall p \in P.$$

Without loss of generality, take $f \in \mathcal{H}(\underline{l}, \bar{l})$ such that $v = u_i \circ f$ holds for all $i \in \{0\} \cup I$, and take $g \in \mathcal{H}(\underline{l}, \bar{l})$ such that $\beta \mathbf{1} = u_i \circ g$ holds for all $i \in \{0\} \cup I$. Then, $\langle u_i \circ f, p_i \rangle < \langle u_i \circ g, p_i \rangle$ holds for all $i \in I$, but we have $\langle u_0 \circ f, p_0 \rangle \geq \langle u_0 \circ g, p_0 \rangle$, which is a violation of Common Taste Pareto. □

11.5 Commutativity Between Ex-Ante Approach and Ex-Post Approach

Let us forget about subjective expected utility theory once, and start with thinking of a "reasonable" method of social decision under uncertainty in general. We will reach the same impossibility.

When we calculate "expected social welfare" under uncertainty, there will be two natural methods. One is to first calculate each individual's expected welfare, which may or may not be the expected utility in the standard sense, by applying a certain decision model, and then aggregate them by applying a certain social welfare criterion. Let us call it the ex-ante approach. The other is to first calculate ex-post social welfare at each state, by applying some social welfare criterion, and calculate the ex-ante evaluation of that by applying some decision model. Let us call it the ex-post approach.

We would think that the two methods should yield the same thing, since we want consistency and do not want to be affected by the order of aggregation. Below, let us see how such a requirement restricts social decision.

Assume that the set of states S is finite and let I be the finite set of individuals. Assume that each individual can receive their *own* outcome at each state and it is given as a real number. Note the individual i's outcome at state s by f_{is}, then the uncertainty faced by the society overall is written in the form of a matrix $f = (f_{is})_{i\in I, s\in S}$. Let us adopt this as the choice object. The set of all such matrices is given by $\mathbb{R}^{I\times S}$, and we define the social ranking $\succsim$ over this space.

In the ex-ante approach, first we determine each individual's ranking over their own uncertain prospects. When $f_i = (f_{is})_{s\in S}$ is at least as desirable as $g_i = (g_{is})_{s\in S}$ for $i \in I$, we write $f_i \succsim_i g_i$. Note that such ranking is for the purpose of socially evaluating individual i's uncertain prospects it may or may not be related to i's own preference.

In order that the ranking $\succsim_i$ is defined for each i consistently with the entire social ranking, the following must be met.

Axiom 11.14 For all $i \in I$, for all f_i, g_i and $h_{-i}, \widetilde{h}_{-i}$,

$$(f_i, h_{-i}) \succsim (g_i, h_{-i}) \iff (f_i, \widetilde{h}_{-i}) \succsim (g_i, \widetilde{h}_{-i}).$$

The condition $(f_i, h_{-i}) \succsim (g_i, h_{-i})$ states that f_i is at least as desirable as g_i for i when the others face uncertain prospects h_{-i}. In order that the ranking $\succsim_i$ is well-defined, the condition must hold independently of the choice of h_{-i}.

In the ex-post approach, first we determine the ex-post social ordering conditional on each state. When $f_s = (f_{is})_{i\in I}$ is at least as desirable as $g_s = (g_{is})_{i\in I}$ at state $s \in S$, we write $f_s \succsim_s g_s$.

In order that such ranking $\succsim_s$ is defined for each s consistently with the entire social ranking, the following must be met.

Axiom 11.15 For all $s \in S$, for all f_s, g_s and $h_{-s}, \widetilde{h}_{-s}$,

$$(f_s, h_{-s}) \succsim (g_s, h_{-s}) \iff (f_s, \widetilde{h}_{-s}) \succsim (g_s, \widetilde{h}_{-s}).$$

The condition $(f_s, h_{-s}) \succsim (g_s, h_{-s})$ states that f_s is at least as desirable as g_s at state s when h_{-s} will occur at states other than s. In order that the ranking $\succsim_s$ is well-defined, the condition must hold independently of the choice of h_{-s}.

The theorem below is due to Mongin and Pivato [81] and Zuber [102].

Theorem 11.10 *A social ranking* $\succsim$ *satisfies both the ex-ante approach and the ex-post approach if and only if there is a collection of functions* $(\phi_{is})_{i\in I, s\in S}$ *such that*

$$f \succsim g \iff \sum_{i\in I}\sum_{s\in S}\phi_{is}(f_{is}) \geq \sum_{i\in I}\sum_{s\in S}\phi_{is}(g_{is})$$

holds for all f, g.

Also, another collection $(\psi_{is})_{i \in I, s \in S}$ represents $\succsim$ in the above form if and only if there is a number $\alpha > 0$ and a collection of numbers $(\beta_{is})_{i \in I, s \in S}$ such that

$$\psi_{is} = \alpha \phi_{is} + \beta_{is}$$

holds for all i and s.

Proof Since $\succsim$ satisfies weak separability across states, there is a collection of conditional rankings $(\succsim_s)_{s \in S}$, and let $(u_s)_{s \in S}$ denote a collection of their (ordinal) representations. Then we can write

$$U(f) = \Phi\left(u_1\begin{pmatrix} f_{11} \\ \vdots \\ f_{|I|1} \end{pmatrix}, \cdots, u_{|S|}\begin{pmatrix} f_{1|S|} \\ \vdots \\ f_{|I||S|} \end{pmatrix}\right).$$

On the other hand, since $\succsim$ satisfies weak separability across individuals, there is a collection of rankings $(\succsim_i)_{i \in I}$, and let $(U_i)_{i \in I}$ denote a collection of their (ordinal) representations. Then we can write

$$U(f) = W\begin{pmatrix} U_1(f_{11}, \cdots, f_{1|S|}) \\ \vdots \\ U_{|I|}(f_{|I|1}, \cdots, f_{|I||S|}) \end{pmatrix}.$$

By combining the above two, we obtain a functional equation

$$W\begin{pmatrix} U_1(f_{11}, \cdots, f_{1|S|}) \\ \vdots \\ U_{|I|}(f_{|I|1}, \cdots, f_{|I||S|}) \end{pmatrix} = \Phi\left(u_1\begin{pmatrix} f_{11} \\ \vdots \\ f_{|I|1} \end{pmatrix}, \cdots, u_{|S|}\begin{pmatrix} f_{1|S|} \\ \vdots \\ f_{|I||S|} \end{pmatrix}\right).$$

By Aczel and Maksa [2], the solution of this functional equation takes the form

$$U(f) = \sum_{i \in I} \sum_{s \in S} \phi_{is}(f_{is}),$$

$$U_i(f) = \sum_{s \in S} \phi_{is}(f_{is}) \quad u_s(f) = \sum_{i \in I} \phi_{is}(f_{is}).$$

□

Thus, ex-ante welfare valuation for each individual must be additive across states, which is obtained as a consequence. Now if we impose Ex-ante Pareto there must be a common prior $p \in \Delta(S)$ and a list of state-independent vNM indices of individuals $\{u_i\}_{i \in I}$ such that

$$\phi_{is} = p_s u_i$$

holds for all $i \in I$ and $s \in S$. Thus, each individual's preference must follow the subjective expected utility theory and everybody has to have the same belief, and this follows as a consequence of the exchangeability of ex-ante/ex-post approaches and the ex-ante Pareto condition. We reach the same type of impossibility again.

Chapter 12
Social Decision Making over Time

Zuber [101], and Jackson and Yariv [63, 64] showed that when people disagree not only on tastes over per-period outcomes but also on time discounting, the only option is dictatorship. Moreover, Jackson and Yariv [63, 64] have shown that even when there is no disagreement on tastes the society must adopt exactly one person's discount factor, which rules out any kind of averaging.

12.1 Aggregation of Discounted Utility Preferences

Let C be the set of social outcomes per period, which is assumed to be finite for simplicity. Also for simplicity of the argument, let us assume that the society can provide a lottery over C at each period. Thus, the set of lottery outcomes per period is $\Delta(C)$, where its generic element is denoted by $l \in \Delta(C)$.

We consider that time is discrete. Let $T \geq 3$ be the time horizon, which may be infinite. Thus the set of streams of social outcomes is $\Delta(C)^T$, where its generic element is denoted by $\mathbf{l} = (l_1, l_2, \cdots) \in \Delta(C)^T$. Note that $\Delta(C)^T$ is endowed with the product metric.

Let I be a finite set of individuals. Each $i \in I$ has a preference $\succsim_i$ over $\Delta(C)^T$, which satisfies the expected discounted utility theory and is represented in the form

$$U_i(\mathbf{l}) = \sum_{t=1}^{T} u_i(l_t)\beta_i^{t-1},$$

where $u_i : \Delta(C) \to \mathbb{R}$ is a mixture-linear function taking the form

$$u_i(l) = \sum_{c \in C} v_i(c)l(c).$$

T. Hayashi, *Decision Theory*, Monographs in Mathematical Economics 9,
https://doi.org/10.1007/978-981-95-2200-2_12

The society is supposed to have a ranking $\succsim_0$ over $\Delta(C)^T$, which satisfies the expected discounted utility theory and is represented in the form

$$U_0(\mathbf{l}) = \sum_{t=1}^{T} u_0(l_t)\beta_0^{t-1},$$

where $u_0 : \Delta(C) \to \mathbb{R}$ is a mixture-linear function taking the form

$$u_0(l) = \sum_{c \in C} v_0(c)l(c).$$

We assume the following minimal agreement condition.

Richness

(i) There exist $\overline{c}, \underline{c} \in C$ such that $\delta(\overline{c})^T \succ_i \delta(\underline{c})^T$ for all $i \in I$, where $\delta(c)^T$ denotes the constant sequence of lotteries that are degenerate on $c \in C$. Without loss of generality, we assume $v_i(\underline{c}) = 0$ for all $i \in I \cup \{0\}$.
(ii) For all $i \in I$ there is $\overline{c} \in C$ such that $\delta(\overline{c}_i)^T \succ_i \delta(\underline{c})^T$ and $\delta(\overline{c}_i)^T \sim_j \delta(\underline{c})^T$ for all $j \neq i$.
(iii) $\beta_1, \cdots, \beta_{|I|}$ are distinct.

The ex-ante Pareto axiom in this context states that if everybody prefers one stream of social outcomes over another so should the society.

Ex-ante Pareto: For all $\mathbf{l}, \mathbf{l}' \in \Delta(C)^T$, if $\mathbf{l} \succ_i \mathbf{l}'$ for all $i \in I$ then $\mathbf{l} \succ_0 \mathbf{l}'$.

Below is the impossibility theorem.

Theorem 12.1 *Assume that $(\succsim_i)_{i \in I \cup \{0\}}$ satisfies the discounted utility theory and Richness. Then it satisfies Ex-ante Pareto if and only if there is $i \in I$ such that $\succsim_0 = \succsim_i$.*

Proof Because of Ex-ante Pareto and Richness (i), a version of the Harsanyi theorem (Harsanyi [55], De Meyer and Mongin [20]) we have linear aggregation

$$U_0(\mathbf{l}) = \sum_{i \in I} \alpha_i U_i(\mathbf{l}).$$

with $\alpha \in \mathbb{R}_+^I \setminus \{\mathbf{0}\}$.

From the discounted utility representation, we have

$$\sum_{t=1}^{T} u_0(l_t)\beta_0^{t-1} = \sum_{i \in I} \alpha_i \sum_{t=1}^{T} u_i(l_t)\beta_i^{t-1}.$$

Now restricting attention to sequences with the form $(l, \delta(\underline{c})^T)$, we obtain

$$u_0(l) = \sum_{i \in I} \alpha_i u_i(l).$$

Also, by restricting attention to sequences with the form $(\delta(\underline{c}), l, \delta(\underline{c})^T)$, we obtain

$$u_0(l)\beta_0 = \sum_{i \in I} \alpha_i u_i(l)\beta_i.$$

By combining the above two equalities, we obtain

$$\left\{\sum_{i \in I} \alpha_i u_i(l)\right\} \beta_0 = \sum_{i \in I} \alpha_i u_i(l)\beta_i,$$

that is,

$$\sum_{i \in I} \alpha_i u_i(l)(\beta_i - \beta_0) = 0.$$

Suppose $\beta_0 \neq \beta_i$ for all $i \in I$. Then for each $i \in I$, by taking $l = \delta(\overline{c}_i)$ we obtain $\alpha_i v_i(\overline{c}_i)(\beta_i - \beta_0) = 0$. Hence $\alpha_i = 0$ for all $i \in I$, which contradicts α being a non-zero vector.

Hence $\beta_0 = \beta_i$ for some $i \in I$. By the same argument, we obtain $\alpha_j = 0$ for all $j \neq i$. □

One can show the above impossibility result even without the assumption of additive discounted utility or the domain with lotteries. Zuber showed that when individuals' preferences and the society's objective function satisfy the stationary ordinal utility theory due to Koopmans and the ex-ante Pareto axiom is met then the only possibility is that the individuals' lifetime utility functions are additive, the society's objective function is additive, and the aggregation rule must be additive and the discounted factors must be identical.

Note that such a "discounting dictator" exists even when there is no disagreement on tastes. Assume Richness (i) and define $D = \{\mu\delta(\overline{c}) + (1 - \mu)\delta(\underline{c}) : \mu \in [0, 1]\}$. Then

$$l^T \succsim_i m^T \iff l^T \succsim_j m^T$$

holds for all $i, j \in I$ and $l, m \in D$.

Common Taste Pareto: For all $\mathbf{l}, \mathbf{l}' \in D^T$, if $\mathbf{l} \succ_i \mathbf{l}'$ for all $i \in I$ then $\mathbf{l} \succ_0 \mathbf{l}'$.

Theorem 12.2 *Assume that $(\succsim_i)_{i\in I\cup\{0\}}$ satisfies the discounted utility theory and Richness (i). Then it satisfies Common Taste Pareto if and only if there is $i \in I$ such that $\beta_0 = \beta_i$.*

Proof Without loss of generality, assume that $u_i = u_0$ holds over D for all $i \in I$. Also, assume $u_0(\delta(\overline{c})) = 1$ and $u_0(\delta(\underline{c})) = 0$, without loss of generality.

Because of Common Taste Pareto and Richness (i), a version of Harsanyi theorem (Harsanyi [55], De Meyer and Mongin [20]) we have linear aggregation

$$U_0(\mathbf{l}) = \sum_{i\in I} \alpha_i U_i(\mathbf{l}),$$

with $\alpha \in \mathbb{R}^I_+ \setminus \{\mathbf{0}\}$.

From the discounted utility representation, we have

$$\sum_{t=1}^{T} u_0(l_t)\beta_0^{t-1} = \sum_{i\in I} \alpha_i \sum_{t=1}^{T} u_0(l_t)\beta_i^{t-1}.$$

Now restricting attention to sequences in which $\delta(\overline{c})$ appears at Period t and $\delta(\underline{c})$ elsewhere, we obtain

$$\beta_0^{t-1} = \sum_{i\in I} \alpha_i \beta_i^t.$$

From the case of $t = 1$, we have $\sum_{i\in I} \alpha_i = 1$. From the case of $t = 2$, we have

$$\beta_0 = \sum_{i\in I} \alpha_i \beta_i.$$

Hence we have

$$\left(\sum_{i\in I} \alpha_i \beta_i\right)^{t-1} = \sum_{i\in I} \alpha_i \beta_i^t.$$

Suppose $\alpha_i > 0$ and $\alpha_j > 0$ for distinct i and j. Then since function $f(\beta) = \beta^{t-1}$ is strictly convex for $t \geq 3$, we have

$$\left(\sum_{i\in I} \alpha_i \beta_i\right)^{t-1} < \sum_{i\in I} \alpha_i \beta_i^t,$$

which is a contradiction to the preceding claim. □

12.2 Dropping Stationarity?

If we simply drop stationarity, for example the weighted utilitarian aggregation of ex-ante lifetime utility

$$\sum_{i\in N}\alpha_i\sum_{t=1}^{T}u_i(l_t)\beta_i^{t-1}$$

satisfies Ex-ante Pareto and indeed it is characterized by Ex-ante Pareto since the same proof for the Harsanyi-type aggregation works. To see that it is not stationary in general, take the projection from Period τ, then we have

$$\sum_{i\in N}\alpha_i\sum_{t=\tau}^{T}u_i(l_t)\beta_i^{t-1}=\sum_{i\in N}\alpha_i\beta_i^{\tau-1}\sum_{t=\tau}^{T}u_i(l_t)\beta_i^{t-\tau},$$

which is renormalized into the form

$$\sum_{i\in N}\frac{\alpha_i\beta_i^{\tau-1}}{\sum_{j\in N}\alpha_j\beta_j^{\tau-1}}\sum_{t=\tau}^{T}u_i(l_t)\beta_i^{t-\tau}.$$

Therefore, all individuals' relative welfare weights other than the most patient one converge to zero in the long-run, unless all individuals are equally patient (which is generically impossible). This is consistent with the classic result that Pareto efficiency implies all individuals' consumption and wealth other than the most patient one converge to zero in the long-run (Becker [10]).

It could still be dynamically consistent, at a logical level, by arguing that stationarity is the synonym of dynamic consistency only under time-invariance (Millner and Heal [78]). We verified this in Chap. 8. In other words, time-*de*pendence means that today we could allow a claim like "today is special, give me a favor," but when today ends and tomorrow comes we are supposed to commit to saying that the special day is indeed over and there will be no more favors, to be dynamically consistent. This is an empirical question about what the society can commit to, but I am skeptical about the assumption that the society can commit to the maximization solution for such a non-stationary objective function, unless there is some commitment device or extra discipline, because when somebody says "today is special, give me a favor" today it is natural to expect that he or she says the same thing tomorrow and repeats it onward. At least, the society continues to face such political pressure every period in the same manner, and there must be a theory about how to deal with it. For a not so successful attempt, see Hayashi [59], which proposes a discipline that if we bring a normative reasoning today to justify something the same reasoning must apply tomorrow and onward.

12.3 Weakening Pareto

Here we present two ways to weaken the Pareto axiom.

12.3.1 *Treating "Successive Selves" of Each Individual as Different Individuals*

Feng and Ke [34] proposed to treat successive selves of each individual as different individuals.

Intergenerational Pareto: For all $\mathbf{l}, \mathbf{l}' \in \Delta(C)^T$, if

$$\sum_{t=\tau}^{T} u_i(l_t)\beta_i^{t-\tau} \geq \sum_{t=\tau}^{T} u_i(l'_t)\beta_i^{t-\tau}$$

for all $i \in I$ and $\tau = 1, \cdots, T$, and

$$\sum_{t=\tau}^{T} u_i(l_t)\beta_i^{t-\tau} > \sum_{t=\tau}^{T} u_i(l'_t)\beta_i^{t-\tau}$$

for at least one $i \in I$ and $\tau = 1, \cdots, T$, then

$$\sum_{t=1}^{T} u_0(l_t)\beta_0^{t-1} > \sum_{t=1}^{T} u_0(l'_t)\beta_0^{t-1}.$$

Here we assume that T is finite.

Theorem 12.3 *Assume that T is finite. Assume that $(\succsim_i)_{i \in I \cup \{0\}}$ satisfies the discounted utility theory and Richness. Then it satisfies Intergenerational Pareto if and only if there exists a vector $\alpha \in \mathbb{R}_+^I \setminus \{\mathbf{0}\}$ such that $u_0 = \sum_{i \in I} \alpha_i u_i$ and $\beta_0 > \max_{i \in I} \beta_i$.*

Proof **"If" Part** Because of Intergenerational Pareto and Richness (i), a version of the Harsanyi theorem (Harsanyi [55], De Meyer and Mongin [20]) we have linear aggregation

$$\sum_{t=1}^{T} u_0(l_t)\beta_0^{t-1} = \sum_{i \in I} \sum_{\tau=1}^{T} \alpha_{i\tau} \sum_{t=\tau}^{T} u_i(l_t)\beta_i^{t-\tau},$$

with $A = (\alpha_{i\tau})$ is an $|I| \times T$ positive matrix.

Now restrict attention to sequences with the form in which l appears at Period 1 and $\delta(\underline{c})^T$ elsewhere, we obtain

$$u_0(l) = \sum_{i \in I} \alpha_{i1} u_i(l).$$

Then restrict attention to sequences with the form in which l appears at Period 2 and $\delta(\underline{c})^T$ elsewhere, we obtain

$$u_0(l)\beta_0 = \sum_{i \in I} (\alpha_{i1} + \alpha_{i2}\beta_i) u_i(l).$$

Combined with the above we obtain

$$\beta_0 \sum_{i \in I} \alpha_{i1} u_i(l_t) = \sum_{i \in I} (\alpha_{i1}\beta_i + \alpha_{i2}) u_i(l).$$

Because of Richness (ii), we can take $(u_i(l))_{i \in I}$ as linearly independent, hence we have

$$\beta_0 \alpha_{i1} = \alpha_{i1}\beta_i + \alpha_{i2}$$

for all $i \in I$. Thus we have

$$\beta_0 = \beta_i + \alpha_{i2}/\alpha_{i1} > \beta_i$$

for all $i \in I$. Now let $\alpha_i = \alpha_{i1}$ for each $i \in I$.

"Only If" Part It suffices to show that there exists a positive sequence $(\widetilde{\alpha}_{i\tau})$ such that

$$\sum_{t=1}^{T} \beta_0^{t-1} \sum_{i \in I} \alpha_i u_i(l) = \sum_{i \in I} \sum_{\tau=1}^{T} \widetilde{\alpha}_{i\tau} \sum_{t=\tau}^{T} u_i(l_t) \beta_i^{t-\tau}.$$

It is done by taking

$$\widetilde{\alpha}_{i1} = \alpha_i$$

and

$$\widetilde{\alpha}_{i\tau} = \alpha_i \beta_0^{t-\tau} (\beta_0 - \beta_i)$$

for all $\tau \geq 2$, for each $i \in I$.

□

Note that the above result says the society must be more patient than any individual even when there is only one individual and even when all individuals agree on discounting. Hence it involves a kind of paternalism.

12.3.2 *Collective Responsibility for Time Discounting*

Here we consider that the individuals are collectively responsible for time discounting. Hence, it allows no paternalism when there is only one individual or when there is no disagreement on time discounting. We follow Hayashi and Lombardi [60] here.

Consensus Pareto: For all $\mathbf{l}, \mathbf{l}' \in \Delta(C)^T$, if

$$\sum_{t=1}^{T} u_i(l_t)\beta_j^{t-1} > \sum_{t=1}^{T} u_i(l_t')\beta_j^{t-1}$$

for all $i, j \in I$ then

$$\sum_{t=1}^{T} u_0(l_t)\beta_0^{t-1} > \sum_{t=1}^{T} u_0(l_t')\beta_0^{t-1}.$$

Theorem 12.4 *Assume that* $(\succsim_i)_{i\in I\cup\{0\}}$ *satisfies the discounted utility theory and Richness. Then it satisfies Consensus Pareto if and only if there is* $s \in I$ *such that* $\beta_0 = \beta_s$*, and there exists a vector* $\alpha \in \mathbb{R}^I_+ \setminus \{\mathbf{0}\}$ *such that* $u_0 = \sum_{i\in I} \alpha_i u_i$.

Proof Because of Consensus Pareto and Richness (i), a version of the Harsanyi theorem (Harsanyi [55], De Meyer and Mongin [20]) we have linear aggregation

$$\sum_{t=1}^{T} u_0(l_t)\beta_0^{t-1} = \sum_{i\in I}\sum_{j\in I} \alpha_{ij} \sum_{t=1}^{T} u_i(l_t)\beta_j^{t-1},$$

where $A = (\alpha_{ij})$ is an $|I| \times |I|$ non-negative and non-zero matrix.

Now restrict attention to sequences with the form in which l appears at Period t and $\delta(\underline{c})^T$ elsewhere, we obtain

$$u_0(l)\beta_0^{t-1} = \sum_{i\in I}\sum_{j\in I} \alpha_{ij} u_i(l)\beta_j^{t-1}.$$

By letting $t = 1$, we obtain

$$u_0(l) = \sum_{i \in I} \sum_{j \in I} \alpha_{ij} u_i(l)$$

and plugging this into the equality above we obtain

$$\beta_0^{t-1} \sum_{i \in I} \sum_{j \in I} \alpha_{ij} u_i(l) = \sum_{i \in I} \sum_{j \in I} \alpha_{ij} u_i(l) \beta_j^{t-1}.$$

Because of Richness (ii), we can take $(u_i(l))_{i \in I}$ as linearly independent, hence we have

$$\beta_0^{t-1} \sum_{j \in I} \alpha_{ij} = \sum_{j \in I} \alpha_{ij} \beta_j^{t-1}.$$

for all $i \in I$.

By letting $t = 2$ we have

$$\beta_0 \sum_{j \in I} \alpha_{ij} = \sum_{j \in I} \alpha_{ij} \beta_j.$$

By plugging this to the preceding equality we obtain

$$\left(\sum_{j \in I} \alpha_{ij} \beta_j \right)^{t-1} = \left(\sum_{j \in I} \alpha_{ij} \right)^{t-2} \sum_{j \in I} \alpha_{ij} \beta_j^{t-1}.$$

By letting $t = 3$ we have

$$\left(\sum_{j \in I} \alpha_{ij} \beta_j \right)^2 = \left(\sum_{j \in I} \alpha_{ij} \right) \sum_{j \in I} \alpha_{ij} \beta_j^2,$$

which is simplified to

$$\sum_{j \in I} \sum_{k \in I \setminus \{j\}} \alpha_{ij} \alpha_{ik} (\beta_j - \beta_k)^2 = 0.$$

Since $\beta_1, \cdots, \beta_{|I|}$ are distinct, we have

$$\alpha_{ij} \alpha_{ik} = 0$$

for all $i, j \in I$ and $k \in I \setminus \{j\}$. Thus, every row of A has at most one non-zero entry. For each $i \in I$, let $j(i)$ be the index for such entry. Then $\beta_0 \sum_{j \in I} \alpha_{ij} = \sum_{j \in I} \alpha_{ij}\beta_j$ yields $\beta_0 \alpha_{ij(i)} = \alpha_{ij(i)}\beta_{j(i)}$, implying $\beta_0 = \beta_{j(i)}$. Since this is true for all $i \in I$, the index $j(i)$ is identical for all i. Name this index as s.

Finally, let $\alpha_i = \sum_{j \in I} \alpha_{ij}$ or each $i \in I$, then we have $u_0 = \sum_{i \in I} \alpha_i u_i$. □

Again we have to select exactly one from the individuals' discount factors, which rules out any type of averaging. This is not bad, however, because we are just selecting a number out of those distributed over the (0, 1)-interval and it can be done nicely. We can choose the median, for example.

In fact, by adding two axioms, anonymity and continuity, we can characterize the generalized median in a variable-profile setting. To formulate the variable-profile setting, consider a function f, a mapping from $(\mathcal{V} \times (0,1))^n$ to $\mathcal{V} \times (0,1)$, where $\mathcal{V}$ denotes the set of von-Neumann/Morgenstern indices such that $\mathcal{V}^n$ satisfies the richness property.

Let σ denote a permutation of N, then σ defines a profile after permutation

$$\sigma\left((v_1, \beta_1), (v_2, \beta_2), \ldots, (v_n, \beta_n)\right) = \Big(\left(v_{\sigma(1)}, \beta_{\sigma(1)}\right), \left(v_{\sigma(2)}, \beta_{\sigma(2)}\right), \ldots, \left(v_{\sigma(n)}, \beta_{\sigma(n)}\right) \Big).$$

Anonymity A social discounting selection rule f is *anonymous* if for every permutation σ on N and every profile of individual decision criteria $((v_1, \beta_1), (v_2, \beta_2), \ldots, (v_n, \beta_n))$ in the domain of f, it holds that

$$f\left((v_1, \beta_1), (v_2, \beta_2), \ldots, (v_n, \beta_n)\right) = f\Big(\left(v_{\sigma(1)}, \beta_{\sigma(1)}\right), \left(v_{\sigma(2)}, \beta_{\sigma(2)}\right), \ldots, \left(v_{\sigma(n)}, \beta_{\sigma(n)}\right) \Big).$$

Continuity A social discounting selection rule f is continuous relative to the Euclidean topology.

Given an integer $k \in \{1, \cdots, n\}$ and n numbers $(\beta_1, \cdots, \beta_n) \in (0,1)^n$, let $k(\beta_1, \cdots, \beta_n)$ denote the k-th largest number out of $(\beta_1, \cdots, \beta_n)$. For example, when $k = 1$ it refers to the largest number, when $k = n$ it refers to the smallest number. Also, when n is odd and $k = \frac{n+1}{2}$ it refers to the median. In this sense we call the selection rule the *generalized median rule* (which is a subclass of Moulin [82]).

Theorem 12.5 *A social discounting selection rule satisfies consensus Pareto, anonymity and continuity if and only if there is an integer* $k \in \{1, \cdots, n\}$ *such that for all profiles* $((v_i, \beta_i)_{i \in N}, (v_0, \beta_0)) \in (\mathcal{V} \times (0,1))^n \times (\mathcal{V} \times (0,1))$ *where* $(v_0, \beta_0) = f\left((v_1, \beta_1), (v_2, \beta_2), \ldots, (v_n, \beta_n)\right)$*, it holds that*

$$v_0 = \frac{1}{n} \sum_{i=1}^{n} v_i$$

and

$$\beta_0 = k(\beta_1, \cdots, \beta_n). \tag{12.1}$$

Proof The "if" part is obvious, hence we prove the "only if" part. Suppose that f satisfies Consensus Pareto, anonymity and continuity. Pick any profile $((v_i, \beta_i)_{i \in N}, (v_0, \beta_0)) \in (\mathcal{V} \times (0, 1))^n \times (\mathcal{V} \times (0, 1))$.

By anonymity, we assume without loss of generality that $\beta_1 \leq \beta_2 \leq \cdots \leq \beta_{n-1} \leq \beta_n$. In addition, by continuity, we assume without loss of generality that $\beta_1 < \beta_2 < \cdots < \beta_{n-1} < \beta_n$.

By restricting attention to sequences of the form $(\ell, \underline{c}, \underline{c}, \underline{c}, \cdots)$, where $\ell \in \mathcal{L}$ is an arbitrary lottery, Theorem 12.4 and anonymity imply that

$$v_0 = \frac{1}{n} \sum_{i=1}^{n} v_i.$$

To establish the generalized median rule, we assume that agents' utilities $v_1, \ldots, v_n$ are linearly independent.

Suppose that $\beta_0 = \beta_s$ for some $s \in N$. By continuity, it always holds that $\beta_0 = \beta_s$. Hence, $k = n - s + 1$, so that

$$\beta_0 = k(\beta_1, \cdots, \beta_n).$$

To complete the proof, we need to show that the choice of k is independent of the choice of individual utilities $v_1, \ldots v_n$. Without loss of generality, we know that all $\beta_1, \cdots, \beta_n$ are distinct. If the choice of k depended on the profile $(v_i)_{i \in N}$, then there would be a jump of the social discount factor from some β_s to another $\beta_{s'}$, which would be a violation of continuity. This completes the proof. □

The social welfare function given by the median discount factor is indeed maximized in a dynamic voting environment *without commitment*. Boylan and McKelvey [13] consider that there is a public good which can be saved and reproduced over time, such as environment, and people vote over how much to save such pubic goods, period by period without commitment, where they are assumed to have an identical per-period utility function. They show that at every period the median saving beats everything else, and it is indeed an optimal growth solution for the agent with the median discount factor.

Bibliography

1. Aczel, J. (ed.): Lectures on Functional Equations and their Applications. Academic Press (1966)
2. Aczel, J., Maksa, G.: Solution of the Rectangular m x n Generalized Bisymmetry Equation and of the Problem of Consistent Aggregation. J. Math. Anal. Appl. **203**(1), 104–126 (1996)
3. Aliprantis, C.D., Border, K.C.: Infinite Dimensional Analysis: A Hitchhiker's Guide. Springer Science & Business Media, Berlin (2006)
4. Allais, M.: The foundations of a positive theory of choice involving risk and a criticism of the postulates and axioms of the American School (1952). Expected Utility Hypotheses and the Allais paradox: Contemporary Discussions of the Decisions under Uncertainty with Allais' rejoinder, pp. 27–145. Springer Netherlands, Dordrecht (1979)
5. Alon, S., Schmeidler, D.: Purely subjective maxmin expected utility. J. Econ. Theory **152**, 382–412 (2014)
6. Anscombe, F.J., Aumann, R.J.: A definition of subjective probability. Ann. Math. Stat. **34**(1), 199–205 (1963)
7. Arrow, K.J.: Aspects of the Theory of Risk-Bearing. Yrjo Jahnsson Lectures, 61 pp. Yrjö Jahnssonin Säätiö, Helsinki (1965)
8. Arrow, K.J.: Essays in the Theory of Risk Bearing. Markham, Chicago (1971)
9. Aumann, R.J.: Agreeing to disagree. Ann. Stat. **4**(6), 1236–1239 (1976)
10. Becker, R.A.: On the long-run steady state in a simple dynamic model of equilibrium with heterogeneous households. Q. J. Econ. **95**(2), 375–382 (1980)
11. Bewley, T.F.: Knightian decision theory. Part I. Decis. Econ. Finance **25**, 79–110 (2002)
12. Billot, A., Qu, X.: Utilitarian aggregation with heterogeneous beliefs. Am. Econ. J. Microecon. **13**(3), 112–123 (2021)
13. Boylan, R.T., McKelvey, R.D.: Voting over economic plans. Am. Econ. Rev. **85**(4), 860–871 (1995)
14. Casadesus-Masanell, R., Klibanoff, P., Ozdenoren, E.: Maxmin expected utility over Savage acts with a set of priors. J. Econ. Theory **92**(1), 35–65 (2000)
15. Chambers, C.P., Hayashi, T.: Preference aggregation under uncertainty: Savage vs. Pareto. Games Econ. Behav. **54**(2), 430–440 (2006)
16. Chambers, C.P., Hayashi, T.: Preference aggregation with incomplete information. Econometrica **82**(2), 589–599 (2014)
17. Chew, S.H., Epstein, L.G.: Recursive utility under uncertainty. In: Equilibrium Theory in Infinite Dimensional Spaces, pp. 352–369. Springer, Berlin, Heidelberg (1991)
18. Chew, S.H., Epstein, L.G., Segal, U.: Mixture symmetry and quadratic utility. Econometrica **59**(1), 139–163 (1991)

T. Hayashi, *Decision Theory*, Monographs in Mathematical Economics 9,
https://doi.org/10.1007/978-981-95-2200-2

19. Danan, E., Gajdos, T., Hill, B., Tallon, J.M.: Robust social decisions. Am. Econ. Rev. **106**(9), 2407–2425 (2016)
20. De Meyer, B., Mongin, P.: A note on affine aggregation. Econ. Lett. **47**(2), 177–183 (1995)
21. Debreu, G.: Topological Methods in Cardinal Utility Theory. Mathematical Methods in the Social Sciences, p. 16. Stanford University Press, Stanford (1959)
22. Debreu, G.: Continuity properties of Paretian utility. Int. Econ. Rev. **5**(3), 285–293 (1964)
23. Dekel, E., Lipman, B.L., Rustichini, A.: Standard state-space models preclude unawareness. Econometrica **66**(1), 159–173 (1998)
24. Dekel, E., Lipman, B.L., Rustichini, A.: Representing preferences with a unique subjective state space. Econometrica **69**(4), 891–934 (2001)
25. Dekel, E., Lipman, B.L., Rustichini, A., Sarver, T.: Representing preferences with a unique subjective state space: a corrigendum 1. Econometrica **75**(2), 591–600 (2007)
26. Diamond, P.A.: Cardinal welfare, individualistic ethics, and interpersonal comparison of utility: Comment. J. Polit. Econ. **75**(5), 765 (1967)
27. Dillenberger, D., Lleras, J.S., Sadowski, P., Takeoka, N.: A theory of subjective learning. J. Econ. Theory **153**, 287–312 (2014)
28. Ellsberg, D.: Risk, ambiguity, and the savage axioms. Q. J. Econ. **75**, 643–669 (1961)
29. Epstein, L.G.: Stationary cardinal utility and optimal growth under uncertainty. J. Econ. Theory **31**(1), 133–152 (1983)
30. Epstein, L.G., Le Breton, M.: Dynamically consistent beliefs must be Bayesian. J. Econ. Theory **61**(1), 1–22 (1993)
31. Epstein, L.G., Schneider, M.: Recursive multiple-priors. J. Econ. Theory **113**(1), 1–31 (2003)
32. Epstein, L.G., Segal, U.: Quadratic social welfare functions. J. Polit. Econ. **100**(4), 691–712 (1992)
33. Epstein, L.G., Zin, S.: Substitution, risk aversion and the temporal behavior of consumption and asset returns: a theoretical framework. Econometrica **57**, 937–969 (1989)
34. Feng, T., Ke, S.: Social discounting and intergenerational Pareto. Econometrica **86**(5), 1537–1567 (2018)
35. Fishburn, P.C.: Utility theory for decision making. No. RAC-R-105. Research analysis corp McLean VA (1970)
36. Fishburn, P.C.: A study of lexicographic expected utility. Manag. Sci. **17**(11), 672–678 (1971)
37. Fudenberg, D., Tirole, J.: Game Theory. MIT Press, Cambridge (1991)
38. Gajdos, T., Hayashi, T., Tallon, J.M., Vergnaud, J.C.: Attitude toward imprecise information. J. Econ. Theory **140**(1), 27–65 (2008)
39. Gayer, G., Gilboa, I., Samuelson, L., Schmeidler, D.: Pareto efficiency with different beliefs. J. Legal Stud. **43**(S2), (2014), S151–S171
40. Geanakoplos, J.: Common Knowledge. Handbook of Game Theory with Economic Applications, vol. 2, pp. 1437–1496. North Holland, Amsterdam (1994)
41. Ghirardato, P.: Revisiting savage in a conditional world. Econ. Theory **20**(1), 83–92 (2002)
42. Ghirardato, P., Marinacci, M.: Ambiguity made precise: A comparative foundation. J. Econ. Theory **102**(2), 251–289 (2002)
43. Ghirardato, P., Maccheroni, F., Marinacci, M., Siniscalchi, M.: A subjective spin on roulette wheels. Econometrica **71**(6), 1897–1908 (2003)
44. Gilboa, I.: Theory of Decision under Uncertainty. Cambridge University Press, Cambridge (2009)
45. Gilboa, I., Schmeidler, D.: Maxmin expected utility with non-unique prior. J. Math. Econ. **18**(2), 141–153 (1989)
46. Gilboa, I., Samet, D., Schmeidler, D.: Utilitarian aggregation of beliefs and tastes. J. Polit. Econ. **112**(4), 932–938 (2004)
47. Gilboa, I., Maccheroni, F., Marinacci, M., Schmeidler, D.: Objective and subjective rationality in a multiple prior model. Econometrica **78**(2), 755–770 (2010)
48. Gorman, W.M.: Community preference fields. Econometrica **21**(1), 63–80 (1953)
49. Gorman, W.M.: The structure of utility functions. Rev. Econ. Stud. **35**(4), 367–390 (1968)

50. Grandmont, J.M.: Continuity properties of a von Neumann-Morgenstern utility. J. Econ. Theory **4**(1), 45–57 (1972)
51. Gul, F., Pesendorfer, W.: Temptation and self-control. Econometrica **69**(6), 1403–1435 (2001)
52. Gul, F., Pesendorfer, W.: Self-control and the theory of consumption. Econometrica **72**(1), 119–158 (2004)
53. Hammond, P.J.: Changing tastes and coherent dynamic choice. Rev. Econ. Stud. **43**(1), 159–173 (1976)
54. Hammond, P.J.: Ex-ante and ex-post welfare optimality under uncertainty. Economica **48**(191), 235–250 (1981)
55. Harsanyi, J.C.: Cardinal welfare, individualistic ethics, and interpersonal comparisons of utility. J. Polit. Econ. **63**(4), 309–321 (1955)
56. Hartmann, L.: Savage's P3 is redundant. Econometrica **88**(1), 203–205 (2020)
57. Hayashi, T.: Intertemporal substitution, risk aversion and ambiguity aversion. Econ. Theory **25**(4), 933–956 (2005)
58. Hayashi, T.: Imprecise information and subjective belief. Int. J. Econ. Theory **8**(1), 101–114 (2012)
59. Hayashi, T.: Consistent updating of social welfare functions. Soc. Choice Welfare **46**(3), 569–608 (2016)
60. Hayashi, T., Lombardi, M.: Social discount rate: spaces for agreement. Econ. Theory Bull. **9**(2), 247–257 (2021)
61. Heifetz, A., Meier, M., Schipper, B.C.: Interactive unawareness. J. Econ. Theory **130**(1), 78–94 (2006)
62. Herstein, I.N., Milnor, J.: An axiomatic approach to measurable utility. Econometrica **21**(2), 291–297 (1953)
63. Jackson, M.O., Yariv, L.: Present bias and collective dynamic choice in the lab. Am. Econ. Rev. **104**(12), 4184–4204 (2014)
64. Jackson, M.O., Yariv, L.: Collective dynamic choice: the necessity of time inconsistency. Am. Econ. J. Microecon. **7**(4), 150–178 (2015)
65. Karni, E., Schmeidler, D.: Atemporal dynamic consistency and expected utility theory. J. Econ. Theory **54**(2), 401–408 (1991)
66. Karni, E., Viero, M.L.: Reverse Bayesianism: a choice-based theory of growing awareness. Am. Econ. Rev. **103**(7), 2790–2810 (2013)
67. Klibanoff, P., Marinacci, M., Mukerji, S.: A smooth model of decision making under ambiguity. Econometrica **73**(6), 1849–1892 (2005)
68. Koopmans, T.C.: Stationary ordinal utility and impatience. Econometrica, 287–309 (1960)
69. Koopmans, T.C.: Representation of preference orderings with independent components of consumption. In: McGuire, R. (eds.) Decision and Organization, pp. 57–78. North-Holland, Amsterdam (1972)
70. Koopmans, T.C.: Representation of preference orderings over time. In: McGuire, R. (eds.) Decision and Organization, pp. 79–100. North-Holland, Amsterdam (1972)
71. Kreps, D.M.: A representation theorem for "preference for flexibility". Econometrica **47**, 565–577 (1979)
72. Kreps, D.: Notes on the Theory of Choice. Routledge, Abingdon (2018)
73. Kreps, D.M., Porteus, E.L.: Temporal resolution of uncertainty and dynamic choice theory. Econometrica **46**, 185–200 (1978)
74. Laibson, D.: Golden eggs and hyperbolic discounting. Q. J. Econ. **112**(2), 443–478 (1997)
75. Li, J.: Information structures with unawareness. J. Econ. Theory **144**(3), 977–993 (2009)
76. Machina, M.J.: Dynamic consistency and non-expected utility models of choice under uncertainty. J. Econ. Lit. **27**(4), 1622–1668 (1989)
77. Machina, M.J., Schmeidler, D.: A more robust definition of subjective probability. Econometrica **60**, 745–780 (1992)
78. Millner, A., Heal, G.: Time consistency and time invariance in collective intertemporal choice. J. Econ. Theory **176**, 158–169 (2018)
79. Mongin, P.: Consistent Bayesian aggregation. J. Econ. Theory **66**(2), 313–351 (1995)

80. Mongin, P.: Spurious unanimity and the Pareto principle. Econ. Philos. **32**(3), 511–532 (2016)
81. Mongin, P., Pivato, M.: Ranking multidimensional alternatives and uncertain prospects. J. Econ. Theory **157**, 146–171 (2015)
82. Moulin, H.: On strategy-proofness and single peakedness. Public Choice **35**(4), 437–455 (1980)
83. Nakamura, Y.: Subjective expected utility with non-additive probabilities on finite state spaces. J. Econ. Theory **51**(2), 346–366 (1990)
84. Noor, J.: Commitment and self-control. J. Econ. Theory **135**(1), 1–34 (2007)
85. Osborne, M.J., Rubinstein, A.: A Course in Game Theory. MIT Press, Cambridge (1994)
86. Phelps, E.S., Pollak, R.A.: On second-best national saving and game-equilibrium growth. Rev. Econ. Stud. **35**(2), 185–199 (1968)
87. Pratt, J.W.: Risk Aversion in the Small and in the Large. Econometrica **32**(1/2), 122–136 (1964)
88. Rao, K.P.S.B., Rao, M.B.: Theory of Charges: A Study of Finitely Additive Measures. Academic Press, New York (1983)
89. Rudin, W.: Principles of Mathematical Analysis, vol. 3. McGraw-hill, New York (1964)
90. Sarin, R., Wakker, P.: A simple axiomatization of nonadditive expected utility. Econometrica **60**, 1255–1272 (1992)
91. Savage, L.J.: The Foundations of Statistics. Courier Corporation, New York (1972)
92. Schmeidler, D.: Subjective probability and expected utility without additivity. Econometrica **57**, 571–587 (1989)
93. Schneider, R.: Equivariant endomorphisms of the space of convex bodies. Trans. Am. Math. Soc. **194**, 53–78 (1974)
94. Seo, K.: Ambiguity and second-order belief. Econometrica **77**(5), 1575–1605 (2009)
95. Strotz, R.H.: Myopia and inconsistency in dynamic utility maximization. Rev. Econ. Stud. **23**(3), 165–180 (1955)
96. Takeoka, N.: Subjective probability over a subjective decision tree. J. Econ. Theory **136**(1), 536–571 (2007)
97. Tversky, A., Kahneman, D.: Rational choice and the framing of decisions. J. Bus. **59**(4), pt 2, S251–S278 (1986)
98. Von Stengel, B.: Closure properties of independence concepts for continuous utilities. Math. Oper. Res. **18**(2), 346–389 (1993)
99. Wakker, P.P.: Additive Representations of Preferences: A New Foundation of Decision Analysis, vol. 4. Springer Science & Business Media, Berlin (2013)
100. Weymark, J.A.: A reconsideration of the Harsanyi-Sen debate on utilitarianism. In: Interpersonal Comparisons of Well-being, p. 255. Cambridge University Press (1991)
101. Zuber, S.: The aggregation of preferences: can we ignore the past? Theory Decis. **70**(3), 367–384 (2011)
102. Zuber, S.: Harsanyi's theorem without the sure-thing principle: on the consistent aggregation of Monotonic Bernoullian and Archimedean preferences. J. Math. Econ. **63**, 78–83 (2016)

Index

T. Hayashi, *Decision Theory*, Monographs in Mathematical Economics 9,
https://doi.org/10.1007/978-981-95-2200-2

The manufacturer's authorised representative in the EU is Springer Nature Customer Service Centre GmbH, Europaplatz 3, 69115 Heidelberg, Germany. If you have any concerns regarding our products, please contact ProductSafety@springernature.com

Printed and bound by CPI Group (UK) Ltd, Croydon, CR0 4YY
23/07/2026
02174953-0001